Maximilian Andrews

Molecular Dynamics of
Monomeric IAPP in Solution

A Study of IAPP in Water at
the Percolation Transition

Anchor Academic
Publishing

Andrews, Maximilian: Molecular Dynamics of Monomeric IAPP in Solution: A Study of IAPP in Water at the Percolation Transition, Hamburg, Anchor Academic Publishing 2014

Buch-ISBN: 978-3-95489-323-2
PDF-eBook-ISBN: 978-3-95489-823-7
Druck/Herstellung: Anchor Academic Publishing, Hamburg, 2014

Bibliografische Information der Deutschen Nationalbibliothek:
Die Deutsche Nationalbibliothek verzeichnet diese Publikation in der Deutschen Nationalbibliografie; detaillierte bibliografische Daten sind im Internet über http://dnb.d-nb.de abrufbar.

Bibliographical Information of the German National Library:
The German National Library lists this publication in the German National Bibliography. Detailed bibliographic data can be found at: http://dnb.d-nb.de

© Anchor Academic Publishing, Imprint der Diplomica Verlag GmbH
Hermannstal 119k, 22119 Hamburg
http://www.diplomica-verlag.de, Hamburg 2014
Printed in Germany

To my lovely wife Eve and my caring Family

Contents

List of Figures

List of Tables

Summary

Conformational properties of the full-length human and rat islet amyloid polypeptide 1–37 (amyloidogenic hIAPP and non-amyloidogenic rIAPP, respectively) were studied at physiological temperatures by MD simulations both for the cysteine (reduced IAPP) and cystine (oxidized IAPP) moieties. After performing a temperature scan from 250 K to 450 K at a 20 K interval, it was found that the two temperatures, 310 K and 330 K, delimit the temperature at which the water percolation transition occurs and were thus chosen for observing the conformational properties of IAPP where the biological activity is highest. In fact, most living organisms have the highest biological activity in a temperature interval that corresponds to a percolation transition, which was calculated for hIAPP at ≈320 K and seems to be independent of the chemical composition of the IAPP variant. At all temperatures studied, IAPP does not adopt a well-defined conformation and is essentially random-coil in solution, although transient helices appear forming along the peptide between residues 8 and 22, particularly in the reduced form. Above the water percolation transition, the reduced hIAPP moiety presents a considerably diminished helical content remaining unstructured, while the natural cystine moiety reaches a rather compact state, presenting a radius of gyration that is almost 10 % smaller than what was measured for the other variants, and is characterized by intrapeptide H-bonds that form many β-bridges in the C-terminal region. This compact conformation presents a short end-to-end distance and seems to form through the formation of β-sheet conformations in the C-terminal region with a minimization of the Y/F distances in a two-step mechanism: the first step taking place when the Y37/F23 distance is ≈1.1 nm, with Y37/F15 subsequently reaching its minimum of ≈0.86 nm. rIAPP, which does not aggregate, also presents transient helical conformations. A particularly stable helix is located in proximity of the C-terminal region, starting from residues L27 and P28. These MD simulations show that P28 in rIAPP influences the secondary structure of IAPP by stabilizing the peptide in helical conformations. When this helix is not present, the peptide presents bends or H-bonded turns at P28 that seem to inhibit the formation of the β-bridges seen in hIAPP. Conversely, hIAPP is highly disordered in the C-terminal region, presenting transient isolated β-strand conformations, particularly at higher temperatures and when the natural disulfide bond is present. Such conformational differences found in these simulations could be responsible for the different aggregational propensities of the two different homologues. In fact, the fragment 30–37, identical in both homologues, is known to aggregate in vitro, hence the overall sequence must be responsible for the amyloidogenicity of hIAPP. The increased helicity in rIAPP induced by the serine-to-proline variation at residue 28 seems to be a plausible inhibitor of its aggregation. The specific position of P28 could be more relevant for inhibiting the aggregation than the intrinsic properties of proline alone; in fact, IAPP in cats, which have been observed to develop diabetes mellitus type II and present islet amyloid deposits, contains a proline residue at position 29.

Another characteristic of the above-mentioned compact state of monomeric oxidized hIAPP is that a particularly reactive conformation found along the "folding" pathway is stabilized by the presence of the disulfide bond. Such conformation presents a short end-to-end distance, allowing the peptide to expose the amyloidogenic sequence $N^{22}FGAIL^{27}$ to neighboring peptides. In the reduced hIAPP moiety, this state does not seem to form for any significant amount of time, proven by the fluctuating end-to-end distance. The mean end-to-end distance is smaller than the calculated value for a random-flight chain, proving both the flexibility of hIAPP and the presence of interactions that bring the peptide to compact conformations. Conversely, owing to the intrinsic rigidity of proline, either rIAPP moiety seems to be too rigid to be able to fold to the short end-to-end distance conformations observed for the oxidized hIAPP moiety, although there are instances in which oxidized rIAPP reaches short end-to-end distances, corresponding to the absence of helices in the P28 region. These conformations possibly occur thanks to the disulfide bond/C-terminus interactions, as seen for hIAPP. Short Y37/L23 distances are also observed in the same time frame of short end-to-end distances as seen with Y37/F23 distances in hIAPP, but since leucine is not aromatic, it is possible that the first step in the "folding" process observed in hIAPP cannot occur in the wild-type rIAPP.

In silico mutations have been applied to the "folded" state obtained in the oxidized hIAPP simulations at 330 K, in order to observe what kind of effect proline has on the conformation. In particular, the S28P substitution induces the formation of a helix in this region and disrupts the compact structure by separating the ends of this particularly stable conformation; in fact, the wild-type oxidized homologue remains compact upon heating up to 390 K.

Thus, in light of the results presented in this study, the collapsed state of the monomeric form was observed after the following three conditions had been met:

1. Presence of the disulfide bond: It was observed that the oxidized polypeptide was more flexible than the reduced counterpart and the short end-to-end distance in IAPP was stablized by its presence.
2. Absence of helical content in the C-terminus region: This allowed this region of the polypeptide to be more flexibile and "fold." P28 seemed to stabilize the highly mobile and unstructured portion of IAPP. Moreover, such helicity seemed to inhibit short end-to-end distances.
3. Presence of aromatic residues: Interactions between aromatic residues, in particular F23, seemingly stabilized one of the first steps in "folding."

Since these results have been obtained for the monomeric form, further studies are necessary to determine whether these three structural characteristics are also relevant for the aggregation propensity of IAPP.

Chapter 1

Introduction

The word *protein* was coined by the Swedish scientist Jöns Berzelius in 1838 to describe a certain class of molecules and their importance. [1,2] In fact, it derives from the Greek word προτεῖος 'of primary importance' which in turn derives from the word προτος 'first.' [3,4] After almost 150 years, one can read the opening sentence of the first chapter of the book on the structure and molecular properties of proteins by Creighton [5]

> Virtually every property that characterizes a living organism is affected by proteins.
> *Proteins: Structure and Molecular Properties*, Creighton (1993)

and only wonder how much there still is to discover, in order to fully understand how these organic molecules, constituting living organisms, function.

What is fascinating about proteins is the multitude of roles they have within living organisms, from enzymatic catalysis to transport and storage, and from functions as complex as biogenesis to being simply structural, just to mention a few primary functions carried out by proteins. In other words, each cell carries out its activities through the expression of its genes by means of its working molecules, i.e., the proteins. How many proteins are encoded by a simple unicellular eukaryote like saccharomyces cerevisiae? The predicted number expressed by this yeast genome is 6200 (as can be found on Table 7.3 in *Molecular Cell Biology* by Lodish et al. (2000)).* But what is even more astonishing is the fact that thousands of primary structures are linear chains consisting of a combination of only twenty amino acids. A protein can thus be considered a word, determined by a sequence of letters of the alphabet that has a meaning. [6]

Once the sequence of amino acids has been found, the adventure begins! The reason is that many times the function of the protein is still unknown. In fact, the primary structure of the object of this study, i.e., islet amyloid polypeptide (IAPP), is known, albeit its biological function remains unclear. Moreover, the functionality of proteins and peptides depends on the native conformation, which for IAPP is also still unknown.

IAPP seems to be involved in the regulation of the glucose metabolism, since it is co-secreted with insulin from pancreatic β-cells. Its physiological role is unclear. Although pancreatic amyloid deposits in the islets of Langerhans have been found in more than

---

*The number of proteins encoded by the human genome is still under debate, ranging from 42 000 genes to 65 000–75 000 genes, as can be found on the Human Project Genome Information page http://www.ornl.gov/sci/techresources/Human_Genome/faq/genenumber.shtml.

95% of the type II diabetes patients, the causal relationship between amyloid formation and the disease is still largely unknown.[7-10] The conditions at which it aggregates are also still unclear; in fact, the human IAPP (hIAPP) sequence in healthy individuals is identical to that found in individuals who suffer from non-insulin-dependent diabetes. On the other hand, other variants, presenting a sequence identity of at least 80% such as rodent IAPP (rIAPP), do not aggregate. Moreover, healthy hIAPP-transgenic mice, which release hIAPP and insulin in a regulated manner, do not present any islet amyloid deposits either. Hence, the primary structure of hIAPP alone is not sufficient to cause amyloid formation. In fact, islet amyloid deposits were found only in mice that presented dysfunctional β-cells. One of the characteristics of the deposits formed by amyloidogenic precursor proteins, such as IAPP, is the proximity of the insoluble deposits to where the protein is produced.[11]

Owing to the complexity of living organisms and all the open questions surrounding them, it would be impossible to determine the biological function of IAPP only through MD simulations. In vitro experiments on hIAPP are particularly demanding, as it forms insoluble aggregates within minutes, compared to other amyloidogenic peptides that take 1 to 3 days, like Aβ, responsible for the amyloid deposits in Alzheimer's Disease. The difficulty lies in the identification of the intermediate states that occur when the peptide undergoes a conformational transition from random coil to an aggregation-prone conformation with increased hydrophobicity.[12] Therefore, in silico investigation of monomeric IAPP conformations in liquid water at physiologically relevant temperatures is possible and should, nevertheless, shed light on the initial steps of aggregation. In order to find the proverbial needle in the haystack, a few points were considered to focus on putative conformational properties that could be responsible for aggregation, i.e., proline. In fact, Westermark et al. have shown that the S28-for-P28 substitution greatly inhibits aggregation.[13] Thus, an atomistic investigation by MD simulations could elucidate how different rodent IAPP is from human IAPP and what characteristics could inhibit aggregation, focusing in particular on, but not limited to, proline. In fact, Green et al. have shown that certain mutations in rIAPP, where residues from the hIAPP sequence are substituted into the rodent sequence, e.g., L23F, form fibrils in vitro.[14] Thus, a parallel comparison between wild-type rIAPP and the in silico rIAPP(L23F) mutant could also give some insight on conformational properties, which can be measured experimentally through Fluorescence Resonance Energy Transfer (FRET).

The answer to the aggregation mystery seems to revolve around the nature of proline, not present in hIAPP, and more precisely around the twenty-eighth residue in the IAPP sequence. In fact, the position of P28 in the primary structure might be the residue that inhibits the aggregation, since cats, which can also develop diabetes mellitus type II accompanied by islet amyloid deposits,[12] present a proline residue in position 29 of the IAPP sequence.

1.1 Islet Amyloid Polypeptide

1.1.1 Diabetes Mellitus Type II

Many degenerative diseases, such as Alzheimer's, Parkinson's, Creutzfeldt-Jakob, diabetes mellitus type II, and several other systematic amyloidoses are related to polypeptide aggregation. Human amyloid polypeptide (hIAPP) forms pancreatic amyloid deposits found in the islets of Langerhans in more than 95% of the type II diabetes patients. The causal relationship between amyloid formation and the disease, however, is still largely unknown.[7-10] These deposits were discovered by Opie at the turn of the twentieth century, when he observed *hyalinosis* in postmortem samples of pancreas of individuals suffering from diabetes.[11] Diabetes mellitus type II (DM2, hereafter), or non-insulin-dependent diabetes, is characterized by an increasing peripheral insulin resistance and secretory dysfunction of β-cells.[7*] The β-cell dysfunction is not clear, but β-cell mass loss does occur. The progressive loss of function of the β-cells can be demonstrated before the clinical pathology of hyperglycemia develops.[†11]

Diabetes mellitus type II has been found to develop spontaneously in cats and monkeys (non-human primates), not only in man. It is quite difficult to establish the relationship between islet amyloid deposition and the three following characteristics of diabetes mellitus type II: increased insulin resistance, onset of hyperglycemia, and β-cell dysfunction. Only through pancreatic biopsies would it be possible to monitor the amyloid formation in relation with the above-mentioned characteristics. Through autopsies, extensive islet amyloid deposits have been found in patients who had severe islet dysfunction, i.e., patients who needed insulin replacement therapy, rather than diet or oral hypoglycemic agents. Hence, the β-cells are insufficient and thus unable to supply an adequate amount of insulin, although the sole cause does not seem to be islet amyloid. In fact, patients with long duration of diabetes mellitus type II have been found to have from prevalence <1% up to 90%, with up to 80% islet mass occupied by amyloid.[‡] The length of the disease is, therefore, unrelated to the severity of it.[12] Moreover, healthy elderly subjects have been found with islet deposits, as is the case for patients with benign insulinoma.[15] Spontaneously developing diabetes mellitus type II has been observed in cats and monkeys, and through longitudinal and cross-sectional studies, it was shown that these models of diabetes present a physiologic syndrome similar to that seen in man, i.e., older age of onset, obesity, impaired glucose tolerance progressing to hyperglycemia, and dependence upon insulin therapy. While the development of the disease is associated with progressive islet amyloid deposit, the same does not hold true for man; in fact, the degree of amyloidosis after many years of DM2 is variable. Owing to the lengthy development of the disease, occurring over years, further investigation through laboratory observations was needed. Islet amyloid did not occur in transgenic mice and rats that express the human IAPP gene alone; in fact, other conditions, including increased transgene expression and obesity, brought about by high-fat feeding and genetically determined obesity, were necessary to observe islet amyloid formation. Many features of DM2 in animals have been demonstrated to be similar in man, but some are very different. Thus, the islet amyloidosis does

---

*The IAPP release by β-cells in diabetes mellitus type I is basically none, as a result of β-cell destruction by an autoimmune condition.[11]

†Amyloid deposits are insoluble proteinaceous accumulations formed by a precursor protein and are normally proximal to the location of production and secretion of the protein. Moreover, each fibril is similar and presents a highly ordered structure, consisting of β-sheets with H-bonding along the length of the fibril and characterized by a "cross-β" X-ray diffraction pattern, even though the size, location, and function of the approximately twenty amyloidogenic precursor proteins are quite different from one another. Fibrils, if examined by electron microscopy, are non-branching structures of indeterminate length having a diameter of 5–10 nm.[11]

‡*Prevalence* indicates the percentage of islets affected, whereas *severity* is the percentage of islet area occupied.

not seem to be the primary causative factor for the onset of diabetes in man, as the results from animal models might have suggested. In fact, more than 50% of the subjects have less than 20% prevalence and less than 10% severity, whereas cross-sectional data show that macaca mulatta present 100% prevalence with >80% severity.[12]

1.1.2   Mutations and Homologues

Not all mutations are deleterious and lead to the death of organisms; in fact, even a perfectly adapted protein undergoes mutations. It is part of evolution. Some mutations of the nucleotide sequence are silent, when the codon mutates into a synonym codon, while others are not. The former are called *silent sites* and the latter are called *replacement sites*, as the amino acid is replaced by a different one, expressed by the newly mutated nucleotide sequence. Such replacements can be deleterious, neutral, or advantageous.[16] When two proteins have a correlated evolution, they are called *homologues*. By comparing the homology between two proteins, one can see which residues are essential for the proteins' proper function. If the residue occupies the same position, it is said to be *invariant* and should remain in the same position for the protein to be functional, whereas if it changes, it could be either *conservatively substituted* or *hypervariable*. The former occurs when two amino acids presenting similar properties occupy the same position, i.e., glutamate and aspartate, whereas the latter is more or less indifferent to the change of a residue in a particular position.[17]

Proline and glycine are often used in mutagenic studies in virtue of their backbone conformational properties. Biological functions in vivo depend on the stability of the folded conformation and can be lost through a mutation that destabilizes the active conformation. In other words, negative observations become significant. An inactive mutant can thus be isolated for further mutagenic studies until the function returns. This process allows the identification of the role of the original residue in the folded and functional protein. This mutation can be random or site-specific. An example could be the substitution of a proline, which presumably terminates a helix, with another residue that can extend the helix. Not only can proline and glycine alter the conformational entropy of the unfolded state, but so can a disulfide bond; the introduction or replacement of one of these elements can perturb the stability of the folded state of the protein. In fact, glycine, proline, and cystine are conserved residues. Large hydrophobic residues are also seldom replaced, whereas acidic and hydrophilic residues are often replaced. Relative frequencies of replacement of the residues that differ in the above-mentioned IAPP variants are listed in Table 1.2 on page 6, whereas all twenty residues can be found in Figure 3.2 of Ref. 5. Normally, most of the mutations do not affect the stability of the folded state since natural selection has most probably already optimized the sequence.[5]

In general, proteins can tolerate the mutation of a single residue without significantly altering the native structure, but the functional properties are much more sensitive to changes. A classical example of this is sickle-cell anemia, where the replacement of a polar glutamate with a nonpolar valine leads to a completely different quaternary structure, producing devastating effects.* Conversely, myoglobin and hemoglobin have only

---

*Actually, the hemoglobin present in the sickle-cell anemia (HbS) is a typical case of Darwinian example, where even one single mutation has led to adaptation of organisms that compete in an environment. In fact, Anthony Allison discovered that heterozygote HbS individuals resisted malaria.[17]

20% of the same sequence, yet share large structural, evolutionary, and functional simi-larities.[2] Thus, with this in mind, discovering which effect the 16% divergence between human and rodent sequences may have on the structural properties could be a rather daunting task.

Interestingly enough, islet amyloid formation in diabetes mellitus type II cannot be related *directly* to any post-translational modification of the peptide or gene mutations that would confer increased amyloidogenicity to the peptide.[12] Although, a missense mutation in the exon 3 of the IAPP gene, reported in 4.1% of the Japanese patients subject to diabetes mellitus type II, seems to lead to an earlier and more severe onset of the disease. The S20G mutation* leads to an in vitro aggregation that allows twofold amyloid at a rate threefold higher than human wild-type gene.[18]

1.1.3 IAPP Properties

Islet amyloid polypeptide (IAPP) is a 37 amino acid peptide, secreted by β-cells, and derives from the precursor proIAPP (89 amino acid peptide), through the same enzymes that convert proinsulin to insulin,† i.e., prohormone convertase 1/3 and 2. Both the IAPP and insulin transcription genes are regulated by glucose or differently regulated by Ca^{2+}, and the secretion of either peptide is closely regulated, i.e., the plasma level of IAPP is 1–15% that of insulin.[11] The role of IAPP seems to be insulin inhibitor, as can be deduced from experiments carried out on IAPP knockout mice. The basal level of circulating glucose and insulin was normal, although males exhibited an increased insulin response to glucose administration and a more rapid glucose disappearance in oral and intravenous glucose tolerance tests. Moreover, body mass in males increased by 20%, which could be determined by increased insulin secretion or an effect of IAPP on food intake.[15]

Human and rat/mouse sequences are compared directly in Table 1.1, with the conser-vatively substituted residues in green and the hypervariable ones in red, whereas those residues found in cat and monkey IAPP that differ from hIAPP are indicated in cyan, if conservatively substituted, and magenta, if hypervariable. The two wild-type islet polypeptide variants of human and rat/mouse are highly conserved, being 84% of the primary structure identical. With the exception of residue 18, the different residues are localized between 20 and 29, which can be seen underlined in the hIAPP primary struc-ture in Table 1.1. Thus, the remaining 16% seem to determine the capability of the peptide to aggregate through β-pleated sheet formation, which has been proven to be amyloidogenic in human and in cat. The most noticeable difference between hIAPP and rIAPP is the presence of proline in positions 25, 28, and 29 (Table 1.1, in red) in

---

*The primary structure of hIAPP, along with other homologues, can be found in Table 1.1 on the next page.

†An interesting note on proinsulin, which may also relate to IAPP, can be made on propeptide size. It can be considered the lower limit of the size of a peptide that can be synthesized on a ribosome and translocated into the ER. Insulin is synthesized as a 110 amino acid peptide called preproinsulin. After removing the signaling protein, it is converted to proinsulin; only during its storage is it cleaved into three parts with the removal of the C-peptide. The two remaining chains, A and B, are connected by two intrapeptide disulfide bonds formed before cleavage. Mature insulin, consisting of 55 amino acids, does not reassemble efficiently without the C-peptide, while proinsulin can refold readily.[5] Hence, unfolding and subsequent inability to refold may be a cause for IAPP aggregation; in fact, such hypothesis may also be supported by the fact that full or partial unfolding, rather than misfolding, seems to be a key step in amyloidogenic diseases.[19]

rIAPP, which most likely does not form β-sheets as a result of the presence of proline residues, normally known as β-sheet breakers.[11] Moreover, residue 23 (also in red) in rIAPP replaces an aromatic residue, phenylalanine, with an aliphatic group, leucine. The other substitutions (in green) are not as drastic, but also present amyloidogenic properties. Residues 18 are both basic, histidine in hIAPP and arginine in rIAPP, while residues 26 are both aliphatic, isoleucine in hIAPP and valine in rIAPP. Green et al. have shown that even though rIAPP is not cytotoxic and does not form fibrils, key single substitutions of the hIAPP into the rIAPP sequence, i.e., R18H, L23F, or V26I, could induce fibril formation in rat IAPP, albeit with low yield.[14]

Table 1.1: *IAPP Primary Structures*

	1	10	20	30
human	KCNTATCAT	QRLANFLVHS	SNNFGAILSS	TNVGSNTY-NH$_2$
rat/mouse	KCNTATCAT	QRLANFLVRS	SNNLGPVLPP	TNVGSNTY-NH$_2$
cat	KCNTATCAT	QRLANFLIRS	SNNLGAILSP	TNVGSNTY-NH$_2$
monkey	KCNTATCAT	QRLANFLVRS	SNNFGTILSS	TNVGSDTY-NH$_2$

Single point mutations in genes can change the amino acid that is expressed, and the resulting relative values can be seen in Table 1.2, although it may not correspond to the actual observed frequencies, where the values with significant discrepancies are written in bold font. Some of the replacements occur often, e.g., Thr/Ala, others seldom, e.g., His/Arg. Replacements involving proline also do not occur much, although the ones that are observed most, i.e., Pro/Ala and Pro/Ser, are those that are found in IAPP (proline properties will be discussed in Section 1.1.3.2). Other residues that are observed more than their expected value, e.g., Ile/Val, Asp/Asn, and Ser/Gly, are also present in IAPP.

Table 1.2: *Relative Frequencies of Amino Acid Replacements*

	Observed Values[a]	Expected Values[b]
Histidine/Arginine	10	8
Isoleucine/Valine	66	18
Phenylalanine/Leucine	17	**41**
Proline/Alanine	35	36
Proline/Serine	27	24
Threonine/Alanine	59	39
Threonine/Proline	7	**28**
Aspartic Acid/Asparagine	**53**	19
Serine/Glycine	**45**	16

[a]Observed replacements in 1572 examples of closely related proteins.[20]
[b]Expected replacements obtained from random single-nucleotide mutations.

The first five replacements seen in Table 1.2 occur in human, rat/mouse, monkey, and cat, while the replacements from the sixth to the eighth occur only in monkey, and the last one is relative to the Japanese human IAPP mutation that has been found in diabetic patients.[18] Although histidine and arginine are sometimes classified as basic amino acids,[1,2] they have different characteristics, so it is not surprising that the relative frequencies with which they replace each other are pretty low; in fact, arginine is almost entirely exposed, i.e., only 1% of the residues are buried by at least 95%, while histidine is slightly less exposed, i.e., 17% of the observed proteins are buried by 95% (Table 6.3, Ref. 5, page 231). While lysine and arginine are positively charged in physiological conditions, histidine can be positively or negatively charged, depending on the environment, in virtue of the imidazole ring and is, thus, a good metal binder and is often found in active sites of proteins.[1,2] In human and rat/mouse IAPP, histidine and arginine are the eighteenth residue in the primary structure, located in a region that presents a transient helix seemingly important for biological function,[21] so it is possible to hypothesize that they also have similar behavior, i.e., as basic residues. The phenylalanine/leucine replacement also does not occur much in virtue of their different characteristics, although both are nonpolar/hydrophobic and pack well in the interior of proteins, with residues buried in at least 45% of the residues.[5] Before discussing the monkey mutations, a quick glance at serine/glycine shows that there are indeed more replacements than those expected. Glycine is so different from all other amino acids, as its side chain is simply H. Moreover, serine can also cap the ends of α-helices thanks to the hydroxyl group in the side chain by forming H-bonds with backbone.[5] A highly observed substitution occurs in monkey IAPP, where asparagine is substituted by aspartic acid, and even though both residues are polar and can form H-bonds, the latter is normally negatively charged in solution, whereas the former is neutral.[1,2] The interesting residue replacement in the primary structure is the one occurring in position twenty-five. First of all, it is the only position in the primary structure, along with residue twenty-eight, in which proline inhibits aggregation, i.e., cat IAPP presents a proline residue in position twenty-nine and is known to aggregate. Second, it has the highest variance, as it presents an uncharged polar residue in monkey, i.e., threonine, proline in rat/mouse, and an aliphatic residue in cat, i.e., alanine. The threonine/alanine replacement value is very high, so one could hypothesize that the structural effect these residues have on proteins is negligible, but since they are both present in the amyloidogenic moieties, and not proline, the flexibility of the polypeptide may determine the amyloidogenicity.

The first fraction (residues 1–20) of both hIAPP and rIAPP seems to have a modest helical propensity, whereas the remaining fraction of the peptide (residues 21–37) seems to be less structured. Moreover, such helicity seems to be required for the biologically active state.[21] In fact, the first twenty residues are either invariant (Table 1.1, in black) or conservatively substituted (Table 1.1, in green), which is also true for monkey and cat IAPP (Table 1.1, cyan). Therefore, one may suppose this sequence is conservatively substituted in order to function properly. On the other hand, the residues in the second half of the peptide, residues 20–29 in particular, are hypervariable and most probably do not influence its biological function.

1.1.3.1   IAPP Aggregation

IAPP is the only peptide found in the amyloid deposits, which occur in the islet. The amyloidogenic form is absolutely necessary for amyloid formation, but there are other factors as well. IAPP is produced by the β-cells, which is the site that is most proximal to the amyloid formation, and its overproduction is not the only condition that can lead to islet amyloid deposit and, thus, to β-cell loss. It seems that a β-cell dysfunction is also necessary for the islet amyloid formation; namely, improperly processed proIAPP, found to form fibrils and present in DM2 islet amyloid deposits. This could occur as a result of the disproportionate release of proinsulin relative to processed insulin. Since this change is present in high-risk individuals prior to the development of the disease and the PC 1/3 and PC2 proteolytic enzymes process both proinsulin and proIAPP, it is possible that proIAPP is improperly converted to IAPP. Therefore, amyloidogenic proIAPP may lead to deposits at the early phases of the islet amyloid deposit formation. This processing, in order to be efficient, needs a tightly regulated environment; in fact, optimal pH and calcium concentrations are necessary for processing proIAPP, as shown by in vitro experiments.[11] DM2 fibrils are found almost exclusively at the extracellular sites in the islets, with small deposits located adjacent to the basement membrane of islet capillaries. The basement membrane could anchor aggregates of IAPP or proIAPP, forming, therefore, a "nucleus" for fibril formation.* In fact, the basement membranes contain heparan sulfate proteoglycans (HSPG), involved in synthetic IAPP fibrillogenesis, and proIAPP has a consensus sequence for HSPG.[12] Jha et al. have shown that proIAPP exhibits a much higher amyloidogenic propensity in the presence of negatively charged membranes than in bulk solvent. However, hIAPP is still much more amyloidogenic than proIAPP. Morphological changes have been observed, although differences in the secondary structures of the aggregated species of hIAPP and proIAPP at the lipid interface are small. Unlike hIAPP, proIAPP forms essentially oligomeric-like structures at the lipid interface.[9] Other studies have also shown morphological changes when IAPP interacts with negatively charged membranes; in fact, Lopes et al. show that the N-terminal part of hIAPP interacts strongly with the negatively charged lipid interface, and the peptide forms ordered fibrillar structures through a two-step conformational transition from a largely α-helical to a β-sheet conformation.[8]

1.1.3.2 Proline

Proline is a special amino acid, as the side chain is bonded to the nitrogen of the amino group forming an imino acid. This tertiary nitrogen cannot form hydrogen bonds, given the absence of $N-H$, and is incompatible with α-helical conformations, if not at the N-terminus. Nevertheless, single proline residues can fit in long α-helices by distorting the local helical geometry. The five-member ring that defines proline is relatively rigid and drastically limits the ϕ angle in the Ramachandran plot to $\approx -60°$, where ϕ is the rotation angle of the peptide unit around the $N-C_\alpha$ bond. The secondary structures assumed by proline are poly(Pro)I, poly(Pro)II, and type I and type II β-turns. Proline residues prefer reverse turns (Ref. 5, Table 6.5, page 256), defined by four residues, of which two are not involved in β-sheets, with an H-bond between residues i and i+3, and proline occupying position i+1. Poly(Pro)I and poly(Pro)II are determined by the conformation of proline, as it can be in either *cis*, in form I, and *trans*, in form II; both of which depend on the solvent, with form II predominating in water, acetic acid, and benzyl alcohol, and form I predominating in propanol and butanol. Conformational changes have been observed to occur upon solvent change. The ϕ angles are $-83°$ and $-78°$ for forms I and II, respectively. Proline also plays an important role in structural fibrous

---

*Nucleation seems to occur through a two-step mechanism, with the nucleation, or seed-forming step, followed by an exponential phase of fibril formation. The cytotoxicity of the amyloid deposits seems to be caused by the initial aggregation steps. Interestingly enough, the cytotoxicity of IAPP is inhibited by Congo Red, while the fibrillogenesis is not.[11]

proteins, like collagen, as it can impart rigidity and stability to the structure. Collagen is characterized by a triple helix similar to poly(Pro)II, with a glycine every three residues, i.e., $(-Gly-Xaa-Yaa-)_n$, with a preponderance of hydroxyproline (Hyp) as Xaa or Yaa, where Hyp forms H-bonds between the hydroxyl group and the amide group of the glycine backbone.[5]

Unlike other amino acids, the peptide unit of proline does not have a partial double bond character to it; in fact, the residue preceding proline is more likely to be in a *cis* conformation than other residues, i.e., a 4:1 ratio favoring the *trans* conformation, as opposed to 1000:1, when comparing proline to other residues, respectively. The residue is also slightly distorted from planarity; in fact, $\Delta\omega = -20°$ to $10°$, compared to $\omega = 0°$ and $180°$ for *cis* and *trans* conformations, respectively. The free-energy barrier associated to a *cis-trans* isomerization is $20.4\,kcal\,mol^{-1}$, making it a slow conversion, i.e., $\tau_{1/2} \approx 20\,min$, which is temperature dependent with the rate increasing by a factor of 3.3 every $10\,°C$ within the normal range. The possibility of assuming a *cis* conformation not sterically hindered also affects the conformational properties such as the radius of gyration and end-to-end distance, illustrated in Section 2.4.3 on page 31.[5]

Another interesting characteristic of proline is its presence in rapid degradable proteins. Such proteins contain one or more "PEST" regions, which are segments of 12–60 residues, in primary structures rich in proline, glutamic acid, serine, and threonine.[5] Whether the high amount of proline residues facilitates the degradation of rIAPP in any way, compared to the serine residues in hIAPP, resulting therefore in a limited in vivo IAPP deposit, is unknown and could be worth investigating.

1.2   Hydration Water

Is there a possible explanation as to why the experimentally measured lag time of hIAPP aggregation drops drastically at approximately $320\,K$, as shown by Kayed et al.?[10] Is it a coincidence that another amyloidogenic peptide, like $A\beta42$,[22] also undergoes a conformational transition at a very close temperature?

There are definitely still many questions evolving around biomolecules and their activity. Pioneering studies by Careri et al. have shown that biomolecules regain their biological activity upon recovering the minimum amount of surrounding water molecules that form an infinite hydration network from an ensemble of small water clusters. This threshold is where water undergoes a quasi-2D percolation transition. One layer of water, or a "monolayer," is sufficient for activity of the biomolecule and is referred to as hydration water. These water molecules are connected by H-bonds of two different types: those that span the system, and those that do not. In other words, the H-bonds of the spanning network wrap the biomolecule completely, without covering it entirely, as there can be water molecules or small clusters of water molecules that are not connected by H-bonds to this network. At low temperature, the dimensionality of network of H-bonded water molecules is quasi-2D, and this network of H-bonds envelopes the biomolecule. Upon heating, this H-bond network of the water molecules decreases until breaking into an ensemble of small clusters.* The process that can describe this is a quasi-2D percola-

---

*A clarifying image of the breakage of the H-bond network can be that of a ball in a net. If the net is intact, the ball moves when one of the knots of the net is pulled, as the H-bond network would behave at lower temperature, i.e., by "pulling" one water molecule everything follows. If the net is weak, pulling

tion transition. Moreover, this transition occurs at biologically relevant temperatures.[23] (Ref. 23, 24, and the references therein, include a complete overview of the percolation transition of hydration water in biosystems.)

Studying the conformational changes of the peptide above and below the percolation transition could shed some light on why faster aggregation was measured by Kayed et al.[10]

1.3 Overview

In aqueous solution, hIAPP has been shown to have an essentially disordered conformation as seen in far UV-CD.[10,25–28] However, it may also assume compact structures[27] and a transient sampling of α-helical conformations has been observed;[21,29] the former has been proven through FRET, and the latter, through NMR spectroscopic studies. The Förster distance between tyrosine and phenylalanine measured for hIAPP in the lag phase of the aggregation process is 12.6 Å, which, if compared to the values obtained through a random walk model,[30] i.e., 30 Å for Y37/F23 and 40 Å for Y37/F15, clearly reveals a structure that is more compact than what is expected for a fully unfolded peptide. The two homologues, human and rat IAPP, when free in solution, show comparable structures; in fact, rIAPP adopts structures similar to hIAPP prefibrillar states.[27] Other studies reveal sampling of α-helical conformations in the central region of the peptide for about 40 % of its length, starting approximately after the tight disulfide bond. In fact, the NMR chemical shifts indicate α-helical propensity from residues 5–19 and their temperature coefficients indicate such a region from residues 7–22.

The 20–29 decapeptides of the different homologues were studied in detail with regards to their aggregation propensity, showing that the S28-to-P28 substitution strongly reduced the amyloidogenicity.[13] Normally, proline residues are both β-sheet and α-helix breakers, but if present as the first element in the helix, they may act as an N-capping residue and can also stabilize helices, even at higher temperatures.[31,32] Other residue substitutions, e.g., rIAPP(L23F), seem to promote aggregation in rIAPP, albeit in low yield.[14] In fact, the fragment 30–37, identical in both homologues, aggregates in vitro. Hence, it is probably the overall sequence that influences the amyloidogenicity of IAPP.[33]

The disulfide bond between residues C2 and C7 also plays an important role. In fact, it has been found experimentally that the presence of this disulfide bond in the peptide also changes the kinetics of aggregation, making the reaction much faster and allowing it to form fibers by secondary nucleation, leaving the structure of the IAPP fiber core intact.[34] Moreover, the disulfide of the cystine seems to stabilize the short end-to-end distance in the oxidized moiety of hIAPP,[35] allowing the formation of aggregation-prone β-sheets.[36]

1.4 Objectives

The most astonishing aspect regarding amyloidogenesis is that many precursor proteins, about twenty, differ not only in primary structure and size, but also in location. The main objective of this study was to observe two very similar polypeptide sequences, being

---

one knot could cause the net to break leaving the ball where it is, or even breaking away from the net itself, i.e., only a water molecule or a small cluster would follow when "pulling" a water molecule.

84% conserved, and pinpoint the different conformational properties of the monomer that may induce or hinder peptide aggregation.

Finding conformational differences of the two monomeric polypeptide homologues in solution could shed light on the underlying mechanism of the aggregation pathway of hIAPP and was thus the focus of this work using MD simulations. The properties studied in this work were the interaction of the aromatic residues of hIAPP and rIAPP, including the mutated in silico variant rIAPP(L23F), the influence of the presence, or absence, of the disulfide bond in both homologues, and the effect of proline, in particular residue 28, on the secondary structure of IAPP. These results are presented in Chapter 5, with an outlook on future work on IAPP presented in Chapter 6.

The conformational properties have been calculated by an ad hoc python program that analyzes GROMACS[37–39] trajectory files. An overview of this program can be found in Chapter 2. Certain parameters, i.e., definitions of H-bonds, which are so important for protein aggregation, and Ramachandran angles for the secondary structure, defined in Chapter 2, were obtained through trial and error, as explained in Chapter 3.

Owing to the difficulty in preparing an initial conformation for an unstructured bio-molecule, a detailed description of how the system was prepared can be found in Chapter 3.

A very helpful tool for the investigation of the proper temperature range to be used is the analysis of the percolation transition of the hydration water surrounding the peptide, localizing therefore a temperature-induced conformational change. Theories on percolation on infinite systems have been developed, but the actual determination of the percolation threshold, especially for finite systems, required extensive study. This tedious work was based on determining which of the many properties of a biomolecule should be measured for locating the percolation transition. Amongst the various monitored properties, the preferred properties are the spanning probability and fractal dimension of the largest cluster. Similar calculations performed on other biomolecules/polypeptides[22,40–42] have also been performed on IAPP, where the break occurs at $\approx$320 K via a quasi-2D percolation transition.[43] A more statistically relevant calculation has since been performed and will be presented in Chapter 4.

For convenience, Ref. 43 is available in Appendix A, with the poster presentations in Appendix B, and the initial and final conformations of the oxidized hIAPP moiety at 330 K[44] are available on request.

Chapter 2

Methods

2.1 Molecular Dynamics Simulation Methods in a Nutshell

The Molecular Dynamics Simulation Method is definitely a very powerful tool for investigating molecular conformations and many other properties. The principles behind it are quite simple and can be explained by Newton's law of motion, with trajectories obtained by solving the renown second law, $F = ma$. In order to apply these laws, there are a few assumptions to be made. The first being, that the motion of electrons are ignored, allowing the system to be treated through classical physics. An obvious limitation in this method is the inability to describe bond cleavage. The bonds are thus treated as springs, described by potentials as simple as Hooke's law for a harmonic oscillator, i.e., $F = -kx$. Second, that the potential is obtained through pair-wise vector summation. The relationship between scalar potential and a conservative force, as seen in the following equation

$$F = -\nabla V(\mathbf{r}), \qquad (2.1)$$

allows a generation of trajectories from a distribution of particles, where the potential is obtained by a pair-wise vector sum between the particles that comprise the system. Hence, from distributions of particles, it is possible to obtain potentials, from which forces can be obtained, and thus accelerations, which after a time δt, lead to new positions. This cycle is then repeated, and repeated, and repeated. Each new position is obtained through integration of the acceleration with respect to time, by means of finite difference methods, with the *Verlet Algorithm* being the most used. The MD simulation is deterministic in a way that the past has an influence on the future of the system, also because the kinetic energy is also taken into account to determine the total energy of the system. This deterministic aspect is useful for determining conformational properties of flexible molecules. Normally, MD simulations can sample *NVE* ensembles, where N, the number of particles in the system, V, the volume, and E, the energy, are all kept constant. Modifications can be made in order to sample from other ensembles, for example the isobaric-isothermal ensemble, where pressure and temperature are kept constant instead of volume and energy, as seen in the microcanonical ensemble (*NVE*). The thermodynamical properties are calculated through an average by the number of time steps.[2,45]

Unfortunately, this holds true only if the time interval is small enough for the force to be constant, and normally this is true when it is smaller than the fastest vibration, which occurs for hydrogen bound to heavy atoms, like oxygen, so the maximum time step is ≈ 0.5 fs. In order to consume less computational time, it is possible to apply constrained dynamics, allowing the time step to increase, because the faster vibrations, like those which involve hydrogen bonded to heavy atoms, are "frozen out" by constraining the bond length to the equilibrium length. The suggested time step, when the molecules are flexible, with rigid bonds, allowing translation, rotation, and torsion is 2 fs.[2,45]

Force fields are the sum of functional forms and parameters. Parametrization is performed to reproduce thermodynamical properties using computer simulation techniques and may include vibrational frequencies, other than parameters to reproduce conformational properties, with the aid of cross-terms. The OPLS force field, i.e., optimized parameters for liquid simulations, has been obtained this way. Unfortunately, there are no absolute force fields, as they have been obtained through a parametrization in order to reproduce a certain property, limiting, therefore, their target of application. An generic functional form can be seen as follows:

$$V = V_{bonds} + V_{angles} + V_{torsion} + V_{Lennard-Jones} + V_{Coulomb}, \tag{2.2}$$

where the first three terms are interactions between bonded atoms, i.e., bond length, bond angle, and torsion angle potentials, respectively, while the last two are relative to nonbonded interactions, i.e., van der Waals potential, most often expressed in the common $6/12$ Lennard-Jones form, and electrostatic potential, approximated by the Coulomb's law, respectively. Actually, there is also a fourth term between bonded atoms related to out-of-plane bending, but this is used to enforce planarity and/or chirality to the modeled molecule by the use of dummy atoms and not always necessary. The last terms are usually the ones that require more time to calculate when obtaining the potential during a simulation step. A possible method to treat long-range interactions, without having to perform a cutoff, is the Ewald Method, which was derived from crystallography due to the periodicity of ions in the unit cells of crystal structures. In order to apply this method for biomolecules, a periodic boundary condition is required, as the charges are placed on a lattice and considered to have infinitely many images in space. The smooth particle-mesh Ewald method allows to lower the aforementioned bottleneck for $\mathcal{O}(N^2)$ to $\mathcal{O}(N\log N)$.[2,45]

Water models are many and can be classified in three main types: simple interaction-points with rigid molecules, flexible molecules, and finally models that take polarization effects into account. SPC/E is the updated model of the SPC, a three-center simple point model with charges exactly balanced on H and O. The van der Waals interactions are calculated with a Lennard-Jones function between the oxygen atoms only.[2,45]

And last but not least, the initial conformation of the sample is very important for the outcome of the experiment, in particular the removal of "hot spots," in which the system presents high-energy interactions that can cause instability in the system. The system must therefore be adequately minimized by means of minimization algorithms.[45] Chapter 3 is entirely dedicated to the preparation of the initial conformation of IAPP.

The pros and cons of Molecular Dynamics Simulation Methods can be summarized by stating that since the motion is continuous, it can be used as a bridge between structures and macroscopic kinetic data, although it is expensive to execute and yields a short time span, requiring a high CPU usage.[2]

2.2   Preparation of Initial Conformations

The polypeptide was initially modeled with MOLDEN v.4.4[46] in an α-helical conformation. In order to create the cystine moiety, it was necessary to bring the two thiol groups
of C2 and C7 within 10 % of the equilibrium bond length, (2.048 ± 0.026) Å.[47] This second step was possible after the rotation around a few arbitrary bonds with SWISS-PDB
VIEWER v.3.7sp5.[48] Such program was also used to create an extended conformation of
the polypeptide by dragging the Ramachandran angles to accepted values near $-180°$ for
ϕ (less than $-130°$) and $180°$ for ψ (greater than $140°$). These structures were simulated
with the Molecular Dynamics suite GROMACS v.3.3.1[37–39] using the OPLS-AA/L force
field (with 2001 amino acid dihedrals).[49,50]

The two hIAPP moieties' molecular weights are 3906.33 Da and 3908.35 Da, respectively 535 atoms for the cystine moiety and 537 atoms for the cysteine moiety. All the
residues, including the termini, have been set at the standard ionization state at a pH
of 7.4 at 25 °C of the individual residues, yielding a net charge of $2\,e$ for the uncapped
C-terminus moiety. If the equilibrium K_a of the ionizable side chains are considered, the
only one which might have a partial charge in aqueous solution is histidine; in fact, if
given the Henderson-Hasselbach equation $pH - pK_a = \log[\text{His}]/[\text{HisH}^+]$ at pH 7.4 and
$pK_a = 6.04$,[51] the concentration ratio is 22.9, which yields a net charge of $0.0436\,e$. This
pK_a is relative to an amino acid in solution, therefore the protonation state can change
according to the conformation of the peptide, but it is possible to approximate it to a
single protonated state, where the hydrogen atom is located on Nε2, as it is the most
favorable hydrogen bonding conformation.[5] As seen in Section 2.1, MD simulations
cannot describe bond cleavage; therefore, ionization states are determined and fixed at
the beginning of the simulation. In other words, the protonation state of the residues is
kept constant, rather than pH.[52]*

2.2.1   In vacuo hIAPP Simulations

Both the α-helical and fully extended (β-strand) conformations of either moiety were
minimized with the L-BFGS algorithm[†][53] and then with the Polak-Ribiere conjugate gra-
dient algorithm[‡][54] with a $\mathbf{F}_{max}$ tolerance of 100 kJ mol^{-1}, followed by an MD simulation
in NVT ensemble of 100 ps in vacuo at 1000 K, with a time step of 2 fs, a 0.9 nm cutoff for
short-range interactions, smooth particle-mesh Ewald (SPME)[55] to treat the long-range
Coulombic interactions, and Berendsen thermostat.[56] The following production phase,
needed to sample random configurations to determine a suitable starting structure,[57] was
performed for an additional 1 ns at the same conditions, albeit using the Nosé-Hoover
thermostat.[58,59]

The polypeptide collapses within 20 ps of equilibration to minimize the charge-charge

---

*At the time of the presentation of Ref. 52 by Donnini et al., histidine protonation states were still
work in progress.

†Limited-memory Broyden-Fletcher-Goldfarb-Shanno algorithm by Nocedal.

‡Two methods were used, since the L-BFGS implementation in GROMACS v.3.3.1 is bugged and does
not allow the use of SPME for long-range interactions. First L-BFGS was used with a switch potential,
then when minimizing by means of other algorithms, i.e., CG, or SD for the solvated system, as well as the
other simulations, the SPME method was used. The insertion of a disulfide bridge for such a small loop is
strenuous on the bond and torsion angles of a peptide and therefore requires a sturdy energy minimization.

interactions between the termini. The mean values of end-to-end distance between the C_α atoms of the first and last residues, referred to as r_{eted}, are (0.62 ± 0.14) nm or less in the four 1 ns in vacuo simulations. This seems to be the only structural parameter that is strongly influenced by the charge-charge interaction in vacuo, compared to the value of at least (1.22 ± 0.07) nm obtained through a 200 ns production run performed at 450 K in SPC/E water.[60] The lack of charge screening in vacuo is seen also by the fact that mean values of the maximum distance between heavy atoms, referred to as L_{max}, and the radius of gyration, R_g, are comparable, but slightly smaller than the ones obtained through the solvated run at 450 K. The same holds true for the standard deviation of the mean of R_g and r_{eted}; in fact, the dielectric screening of the medium reduces the long-range Coulomb interactions, allowing the peptide more movement.

2.2.2   Solvated Uncapped hIAPP

MD simulations have been carried out on four additional conformations* per hIAPP moiety obtained in vacuo, along with the above-mentioned initial α-helical conformation as described in Section 2.2. The trajectories on these solvated peptides were compared in order to ensure an unbiased starting conformation to use for the production phase (see Section 2.2.3) and are studied in detail in Chapter 3.

The peptides were solvated using equilibrated SPC/E water.[60] The initial conformations were appropriately minimized and subsequently temperature pre-equilibrated by a short 50 ps NVT run with restraints on the solute and short time steps (0.5 fs) using Berendsen thermostat.[56] Also a short NPT density pre-equilibration of 100 ps using Berendsen thermostat and pressure coupling[56] was carried out before running the equilibration and production runs, using Parrinello-Rahman pressure coupling[61,62] and the Nosé-Hoover thermostat,[58,59] with time constant for both couplings set at 2.0 ps, and a time step of 2 fs collecting data every 2 ps. It is standard procedure to equilibrate the system through a two-step equilibration, first at constant volume followed by a simulation at constant pressure. The preferred choice is the Berendsen thermostat, as this particular thermostat scales the velocities, thus bringing the temperature quickly to equilibrated values.† If a real NVT ensemble is needed, correct fluctuations are obtained by applying the Nosé-Hoover thermostat.[58,59] The same holds true for pressure coupling, i.e., if thermodynamic properties need to be calculated through MD simulations, the Parrinello-Rahman barostat[61,62] needs to be applied.

Constraints were applied to the water molecules by using the SETTLE[63] algorithm, while SHAKE[64] was applied to covalent bonds of the peptide involving hydrogen. Long-range electrostatic interactions were treated using smooth particle-mesh Ewald,[55,65] with short-range interaction cutoffs set at 0.9 nm. Periodic boundary conditions were set in all three directions, with the box size set at 6 nm for the random conformations taken from Section 2.2.1 and 7 nm for the α-helical conformation taken as reference.

The system charge was neutralized by scaling the partial charges on the peptide to neutrality as described in Section 2.3.

---

*Details on how these initial conformations were chosen are discussed in Section 3.1.1.2.

†An interesting tutorial can be found at http://www.bevanlab.biochem.vt.edu/ Pages/Personal/justin/gmx-tutorials/lysozyme/index.html.

2.2.3 Solvated Amide Capped hIAPP

The protonation states are the same as described in Section 2.2.2, with the exception of the C-terminus being amide capped, yielding a net charge of $3\,e$ for hIAPP and $4\,e$ for rIAPP. In order to neutralize the system in solution, the total charge on the biopolymer was also scaled down to neutrality by distributing an equal and opposite charge on the peptide itself, as seen for the uncapped polypeptide.

The isobaric-isothermal MD simulation production runs of 500 ns for each moiety were performed at 1 bar at 310 K and 330 K. These runs were performed on random starting conformations, i.e., conformations that were obtained after an arbitrary pre-equilibration time of at least 50 ns and that presented a $C\alpha$ RMSD of at least 1.23 nm from the initial modeled α-helix, as can be seen in Figure 2.1. These initial pre-equilibration data were discarded to ensure a completely random starting conformation due to the long autocorrelation times of H-bonds and secondary structure at the lower temperatures.

The peptides were solvated using equilibrated SPC/E water.[60] After proper minimization and equilibration of the system, the following 500 ns *NPT* production runs, in which data were collected every 2.0 ps, were performed using the Nosé-Hoover thermostat[58,59] and the Parrinello-Rahman pressure coupling[61,62] with coupling times of 2.0 ps. In order to avoid "hot solvent and cold solute," the solvent and solute were coupled to two different thermostats and barostats. Constraints were applied to the water molecules by using the SETTLE[63] algorithm, while SHAKE[64] was applied to the covalent bonds of the peptide involving hydrogen. Long-range electrostatic interactions were treated using smooth particle-mesh Ewald,[55,65] with short-range interaction cutoffs set at 0.9 nm. Periodic boundary conditions were set in all three directions, with no interaction between adjacent images as the box size was set at 7 nm with the maximum distance between heavy atoms, L_{max}, being no larger than 5.5 nm.

2.3 Scaling Charges

One of the underlying principles of force fields is that the effective energy potentials are additive, i.e., interaction between atoms are described by a functional form, which is the sum of local terms, between bonded atoms, and nonlocal terms, between nonbonded atoms, as described by Eq. (2.2). Deviations from the equilibrium bond length, bond angle, torsion angles, and the Coulomb and van der Waals interactions between atom pairs describe the potential energy of the system.[2] Therefore, the Coulomb term is calculated independently from the other terms, allowing the possibility of slightly modifying the partial charges on the peptide without drastically perturbing the other terms that describe the potential energy of the system.

Oleinikova et al. and Brovchenko et al. in studies on the hydration shell of Lysozyme[66] and $A\beta_{42}$[67] neutralized the charge of the system by scaling the charges, so the same method was chosen in order to compare the results of the hydration shell analysis of these systems without introducing unknowns to the system, such as counter ions, which could noticeably affect the hydration water. In order to neutralize the system in solution, the total charge of the polypeptide, q_i^o, was scaled down to neutrality, q_i^s, by subtracting the partial scaled charge, q_i^p, with the appropriate sign given by the ratio $q_t/|q_t|$ from the

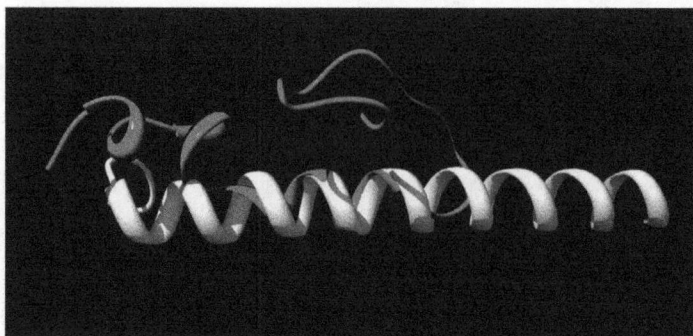

(a) *Red. hIAPP 310 K – initial vs α-helix*

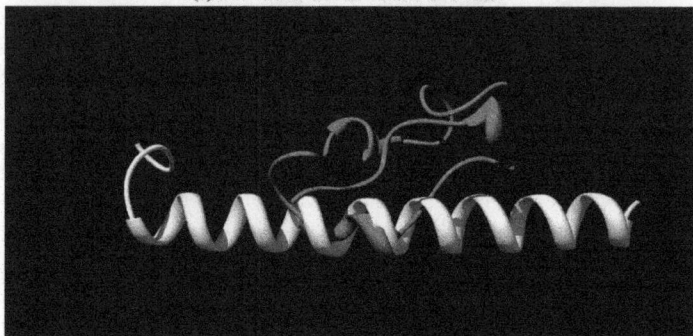

(b) *Red. hIAPP 310 K – after 10 ns vs α-helix*

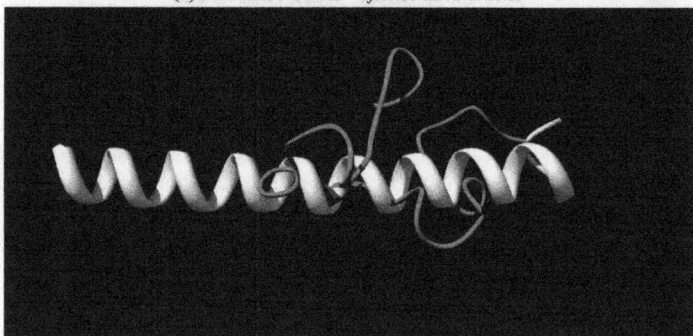

(c) *Ox. hIAPP 310 K – initial vs α-helix*

Figure 2.1: *Comparing hIAPP initial conformations. The white ribbons show the initial α-helical conformation that was simulated at 350 K as described in Section 2.2, while the magenta ribbons show the random conformation obtained after hundreds of nanoseconds, used as initial conformations for the 500 ns production run at 310 K and 330 K analyzed in Chapter 5.*

initial partial charge, q_i^o, as can be seen in the following equations:

$$q_t^o = \sum_{i=1}^{n} q_i^o = +2e, \qquad (2.3)$$

$$q_t^s = \sum_{i=1}^{n} q_i^s = 0, \qquad (2.4)$$

$$q_i^s = q_i^o - \frac{q_t}{|q_t|}|q_i^p|, \qquad (2.5)$$

where the partial charge needed for the charge scaling calculation, q_i^p, is obtained by multiplying the total charge by the contribution of each atom i to the absolute total charge given by the ratio $q_i^o / \sum_{i=1}^{n} |q_i^o|$, as can be seen in the following equation:

$$q_i^p = q_t \frac{q_i^o}{\sum_{i=1}^{n} |q_i^o|}. \qquad (2.6)$$

The scaled partial charges on each atom differ less than 1.5% from the starting value, respectively 1.48% for the cystine moiety and 1.47% for the cysteine moiety,* which leads to an error of less than 3%.†,‡ Moreover, the error that may be introduced by the use of scaled charges is still negligible considering the limitations force fields have in reproducing secondary structures, because of the difficulty in parametrizing the backbone ϕ and ψ dihedral terms.[71] The overall charge of the polypeptide chains is positive, therefore the scaling of the charges makes the negative charges slightly more negative, and the positive charges, a little less positive. Same charge repulsive interactions will be higher in the case of negative charges and lower for positive charges. It is highly unlikely that the secondary structure may be influenced by such a slight change in electrostatic potential, since the difference in interaction between scaled charges, relative to the original unscaled charges, should be negligible compared to the forces involved with the nonbonded interactions governing the secondary structure. In fact, considering a Bland-Altman plot[72] of the Coulomb potential comparing first the effect of different charges on the same structure, and then between independent runs with the different neutralizing methods, it seems the uncertainty introduced is negligible. The system charge was neutralized in the three following ways: scaled charges as previously described (SCAL); a neutralizing charge distributed on the smooth particle-mesh Ewald grid (SPME);[55,65] and a 150 mM sodium chloride concentration to neutralize the charge (NACL), obtained by adding 33 chloride anions and 31 sodium cations randomly.

The Bland-Altman plot[72] is normally used in medicine to test the reliability of new clinical measurements compared to old ones. Calculating the correlation coefficient is

---

*Moreover, none of these charges pass the limit of q_{max}, which is used to define the hydrophobicity in g_sas,[68] so the determination of the solvent accessible surface area is not biased (see GROMACS manual for details).[69]

†The electrostatic potential between two charges q_i and q_j separated by a distance r_{ij} is given by the following relation: $V_{ij} = \frac{q_i q_j}{4\pi\varepsilon_0 r_{ij}}$, with the error given by $\frac{\delta V_{ij}}{V_{ij}} = \sqrt{\left(\frac{\delta q_i}{q_i}\right)^2 + \left(\frac{\delta q_j}{q_j}\right)^2 + \left(\frac{\delta r_{ij}}{r_{ij}}\right)^2} \leqslant \frac{\delta q_i}{q_i} + \frac{\delta q_j}{q_j} + \frac{\delta r_{ij}}{r_{ij}}$.[70]

‡It is possible to run the simulation without scaling the charges and including a reaction field in the Coulomb potential term or by introducing a neutralizing charge on the SPME grid, but it was preferred to scale the charges since the effects it would have on the water model used are still unclear.

not enough, as it may be misleading; in fact, two methods that have been studied to measure the same quantity should be highly correlated. Thus, two independent methods are
compared by graphical interpretation of the difference of the measured quantities plotted
against the average between the two, as can be seen in the right panels of Figures 2.2 and
2.3, where the left panels show the correlation between the same sets of data. The correlation coefficient, r_{xy}, in Figure 2.2 is between 0.97 to 0.99, but that is not surprising, since
the electrostatic potential is calculated for the trajectory with the same (V_S) charges used
for the 30 ns MD simulation runs, then the potential is recalculated with the other charges
(V_O), e.g., for the scaled charges run (SCAL) the electrostatic potential is first calculated
with the scaled charges (same charges of the simulation, indicated with S), then with the
original unscaled charges (other charges, indicated with O). The electrostatic potential between two charges q_i and q_j separated by a distance r_{ij} is given by the following equation:

$$V_{ij} = f \frac{q_i q_j}{r_{ij}}, \qquad (2.7)$$

where f is the **electric conversion factor** equal to $138.935485 \, kJ \, mol^{-1} nm e^{-2}$.[69] The
data on the abscissae in the right panels of Figure 2.2 are the mean value between the two
different electrostatic potentials just mentioned, i.e., $\bar{V}_{S,O} = \frac{V_S + V_O}{2}$, while the ordinates
show the difference between them, $\Delta V_{S,O} = V_S - V_O$.* If the data were truly independent,
the difference between the two sets would be zero, which is not the case. The data have
a systematic error, and this is a way to quantify it. As can be seen by the distribution of
the data and the confidence level of 95 %, or $\pm 1.96\sigma$ (shown in Figures 2.2a and 2.2b),
the data seem to be normally distributed, with 94.2 % and 95.3 % data within the 95 %
confidence level, for the oxidized and reduced hIAPP respectively. The relative mean
value of the difference of electrostatic potential percentage is -1.13% for the cystine
moiety and -1.05% for the cysteine moiety. The other trajectories, i.e., the SPME and
NACL, which have been calculated with the original OPLS-AA partial charges,[49,50]
when substituted with the scaled charges, give an even smaller absolute value of $\Delta V_{S,O}$,
as can be seen in Figures 2.2c through 2.2f.

It is obvious that a systematic error has been introduced, but the question is how large?
The initial estimate was maximum 3 %, around 2 % if calculated by the square root of
the sum of the squares of the relative errors (as seen in Footnote † on the previous page).
The absolute value of the relative error on the calculated electrostatic potential with two
different charge sets on the same conformations is less than 1.13 %. The next step is to
look at the effect the scaled charges have on the electrostatic potential, V_D, compared
to opposite partial charges distributed on the SPME grid, V_G, and neutralizing charges
given by an NaCl solution at physiological ionic strength 150 mM, V_P. As can be seen in
the left panels in Figure 2.3, the correlation between the electrostatic potential of parallel
runs is uncorrelated, with r_{xy} between 0.35 and -0.17, so the Bland-Altman plots in the
left panels should also be uncorrelated. Unfortunately, only comparison of the neutralizing methods with the original charges are uncorrelated, as seen in Figures 2.3e and 2.3f,
where the relative difference between V_G and V_P is -0.2% and 0.1 %, respectively for
oxidized and reduced hIAPP. The scaled charges seem to overestimate the electrostatic
potential. In fact, just like in Figures 2.2a and 2.2b, $\bar{d}$ in Figures 2.3a through 2.3d ranges

---

*In order to make it easier to quantify the differences, $\Delta V_{S,O}$, are divided by the mean values, $\bar{V}_{S,O}$, and
multiplied by 100.

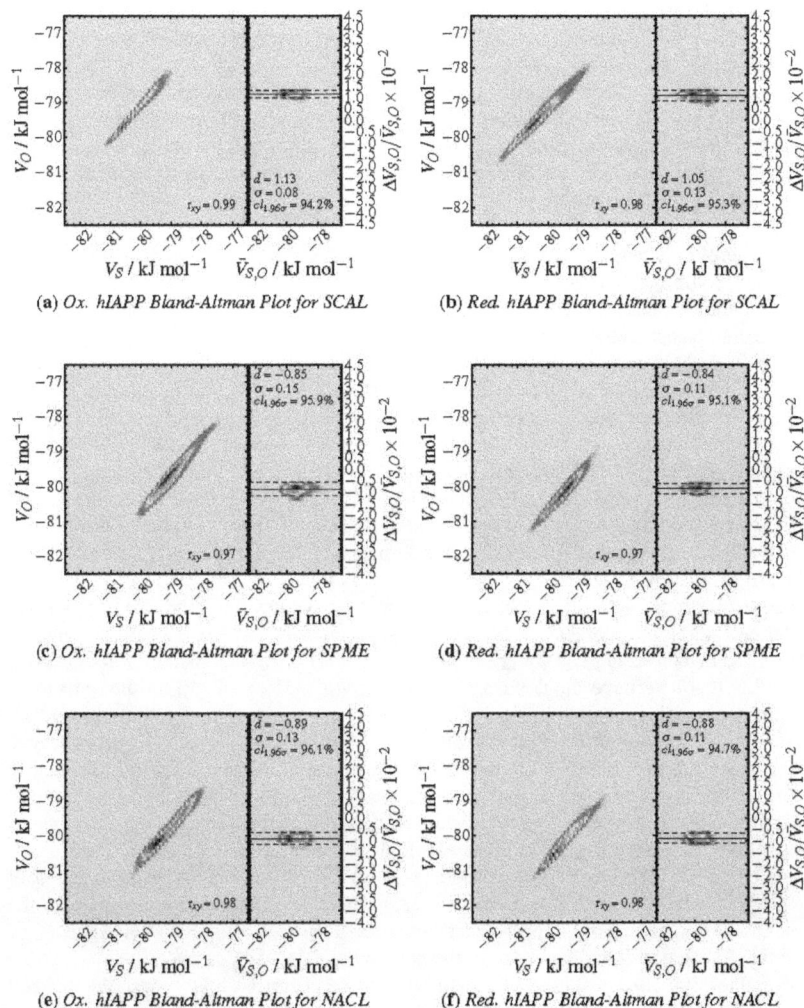

(a) *Ox. hIAPP Bland-Altman Plot for SCAL*

(b) *Red. hIAPP Bland-Altman Plot for SCAL*

(c) *Ox. hIAPP Bland-Altman Plot for SPME*

(d) *Red. hIAPP Bland-Altman Plot for SPME*

(e) *Ox. hIAPP Bland-Altman Plot for NACL*

(f) *Red. hIAPP Bland-Altman Plot for NACL*

Figure 2.2: *Each subfigure shows the correlation between two data sets in the left panels and in the right panels the Bland-Altman plot[72] of the same sets. The occurrence of the data is represented by a reverse spectral color code (ROYGBIV), where the most occurring events are violet and red the least, with black being 100% and gray 0%. Scaled charges for (a) oxidized and (b) reduced hIAPP. Neutralizing charge distributed on SPME grid for (c) oxidized and (d) reduced hIAPP. Physiological 150 mM ionic force for (e) oxidized and (f) reduced hIAPP.*

from 0.9% to 1.4%. This is possibly due to the fact that the unscaled charges belong to neutral *charge groups*, which are no longer neutral after charge scaling. As stated in the manual, if, for example, an atom-atom interaction calculated with O of a water molecule is calculated without the neutralizing charges of the other atoms in the *charge group*, e.g., the two H atoms, a large dipole can be induced in the system.[69] Therefore, in order to avoid this problem, the atom-atom interactions are calculated with all the atoms included in a *charge group*. In the case of the scaled charges, the overall charge of the peptide is neutral; not the charge of each individual group. The atoms still belong to *charge groups*, but the groups may deviate from neutrality because of the scaling of the charge, thus introducing a systematic error.

Therefore, the scaling of the charges can influence the electrostatic potential, which in turn could bias the 1-4 interactions that define the ϕ and ψ torsion angles, which define the secondary structure. As a first estimate, an error of 1.4% in the electrostatic potential, as estimated in Figure 2.3, could correspond to a 5° error in the dihedral angles on the Ramachandran plot, if there were no other forces or barriers involved. This should not be the case, since most torsional terms in OPLS-AA force fields are calculated from ab initio calculations on models using an HF/6-31G* basis set[73] and, thus, should not be influenced by the scaling of the charges. This was investigated by plotting the regions relative to α-helix, β-strands, and poly(L-proline), and a cutoff region 60°×60°, comprising the regions in the Ramachandran plot that contain the maximum peaks of the corresponding areas relative to the considered secondary structures, as can be seen delimited by the dashed red squares in the images on the left of Figures 2.4 to 2.6 and given in detail in Section 2.4.1. The mean values of the data that determine these peaks are at maximum within 5°; moreover, all the mean values lie within the contour that defines the highest content of the secondary structure in consideration. The peaks of these Ramachandran plots are given by the sum of all three runs, with the areas of each marker that determine the contribution of each run to the peak within the red square. None of the charge-neutralizing methods seems to contribute to the peaks more than another, if not the salt solution of reduced hIAPP, NACL in Figure 2.5c. In fact, with the exception of Figure 2.7f, in which there seems to be a more significant content of β-strands than in the other runs, the contribution to the secondary structures for the independent 30 ns trajectories does not differ significantly. If the ϕ and ψ angle acceptance for these secondary structures is increased by 10° in all directions delimiting a cutoff region of 80°×80° (right plots of Figures 2.4 to 2.6), the mean values of ϕ and ψ of the three runs diverge slightly in some runs, with some of the points drifting out of the contour with the maximum occurrence, as can be noticeably seen between Figures 2.5c and 2.5d. Hence, this divergence of the mean points depends on many factors, among them the cutoff, and cannot be solely attributed to the charge scaling. In fact, the standard deviation of the mean, indicating the fluctuation of the system while the conformations project their movement on this plane, is normally 15° when considering the 60°×60° cutoff and reaches values of 20° for the extended cutoff. Another possible control to verify the influence of the charge scaling could have been calculating the dipole moment of the peptide bond H−N−C=O, but unfortunately the GROMACS charge groups differ from this, including also the Cα and Hα, which results in a slightly larger dipole (3.98 D vs. 3.5 D). The neutralizing charge is distributed throughout the entire peptide, so neither charge distribution for the peptide bond is zero, making the calculation of the dipole

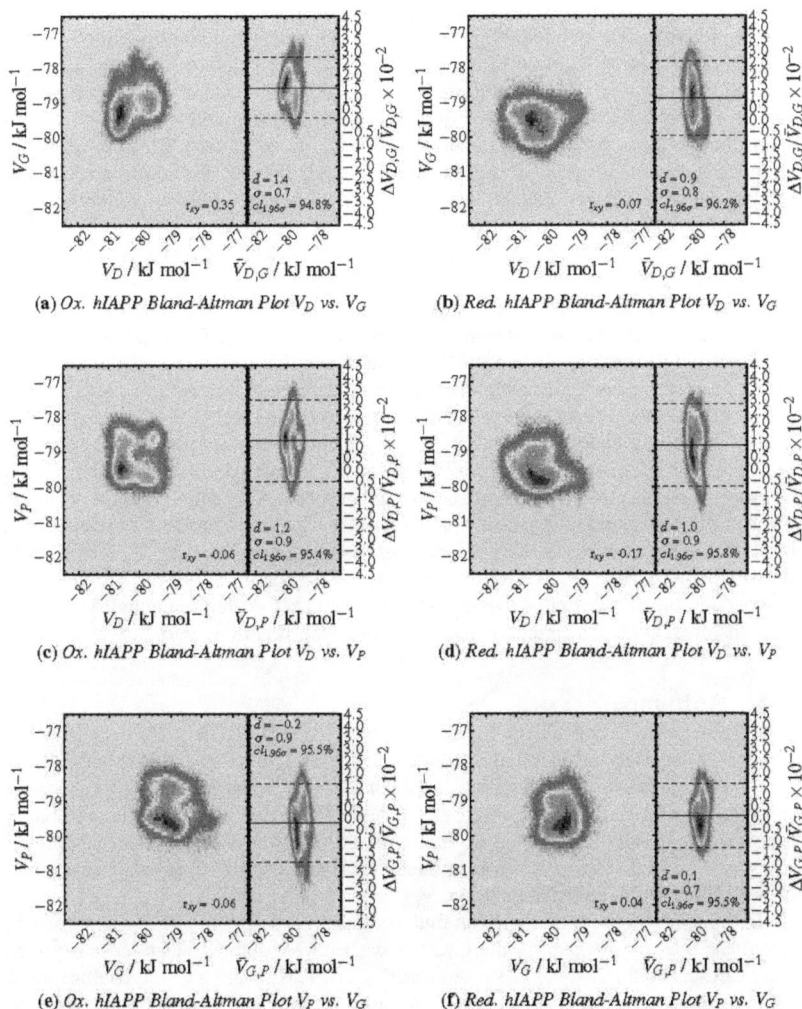

(a) *Ox. hIAPP Bland-Altman Plot V_D vs. V_G*

(b) *Red. hIAPP Bland-Altman Plot V_D vs. V_G*

(c) *Ox. hIAPP Bland-Altman Plot V_D vs. V_P*

(d) *Red. hIAPP Bland-Altman Plot V_D vs. V_P*

(e) *Ox. hIAPP Bland-Altman Plot V_P vs. V_G*

(f) *Red. hIAPP Bland-Altman Plot V_P vs. V_G*

Figure 2.3: *Each subfigure shows the correlation between two data sets in the left panels and in the right panels the Bland-Altman[72] plot of the same sets. The occurrence of the data is represented by a reverse spectral color code (ROYGBIV), where the most occurring events are violet and red the least, with black being 100% and gray 0%. Scaled charges for (a) oxidized and (b) reduced hIAPP. Neutralizing charge distributed on SPME grid for (c) oxidized and (d) reduced hIAPP. Physiological 150 mM ionic force for (e) oxidized and (f) reduced hIAPP.*

23

pointless for comparison since it depends on the Cartesian coordinates of the atoms.* A systematic error in the calculation of the dipole of the peptide bond could influence the overall secondary structure, but it seems that there is no significant difference between the characteristic Ramachandran plots for helical and extended conformations seen in Figure 2.7 on page 28. Because of this uncertainty in the calculation of the dipole, scaled charges should not be used to calculate IR-spectra, since it depends on the variation of the dipole moment,[2] but it should be irrelevant for MD simulations.

Although these 30 ns MD simulation runs at 350 K and 1 bar (*NPT*) are relatively short for statistical purposes, it is possible to conclude that the difference in secondary structure maxima may or may not be induced by the scaling of the charges, but it is certain that the fluctuation of the system is preponderant over any slight effect the scaling of the charges may induce. In fact, such scattering is in line with the scattering of different independent simulation runs.

These data were obtained with uncapped IAPP fearing that the systematic error introduced was larger than with the amide capped system, since this polypeptide is much smaller than Lysozyme as studied by Oleinikova et al.,[66] but the formation of salt bridges tend to bias the trajectories causing short distances between charged groups, whether they are the charged termini end-to-end distances or the charged carboxyl group and R18 in rIAPP. Hence, subsequent runs with amide capped C-termini were performed, where the partial charges for hIAPP are scaled by 2.2 % and 3.0 % for rIAPP. A first approximation of the error on the electrostatic potential for the capped rIAPP, which presents a charge of 4 e, is 4.2 % (as seen in Footnote † on page 19),[70] but as seen for the uncapped hIAPP, it might also be as small as the percentage of the charge scaling, i.e., 2.2 % and 3.0 % for hIAPP and rIAPP, respectively. Moreover, Aβ42, which is a 42 residue 627 atom polypeptide bearing a similar size of IAPP, has also been scaled to neutrality from 3 e without apparent effects on the force field to reproduce system properties.[67]

2.4 Software

The data were analyzed by an ad hoc python program, which uses PYMACS v.0.2[76] for the manipulation of the .xtc files, DSSPcont v.1.0[77,78] and SEGNO v.3.1[79] for the assigning secondary structure, and g_sas[68] of the GROMACS package for calculating the solvent accessible surface area. The error estimation was carried out by applying g_analyze,[80] discussed in detail in Section 2.4.4.1, along with the statistical inefficiency present in MD simulations.

In the following sections, one can find a complete description of various properties calculated by xTc_Rex.py,† along with how they were defined. This script is principally hard-coded for analysis of various forms of IAPP, with and without disulfide bond, i.e., the calculation of aromatic-aromatic distances, the isomeric form of proline in rIAPP,

---

*The central multipole expansion is based upon electric moments, which are the charge, dipole, quadrupole, and so on, and can be represented by a distribution of charges. The first is a scalar, the second a vector, and the third a tensor. Only the first non-zero moment is independent of the coordinates,[45] therefore a dipole of a charged molecule depends on the choice of the origin.

†The name of the ad hoc python script written for the data analysis was coined from its function of analyzing .xtc files and from the bugs that were the size of dinosaurs, i.e., T-Rex. After a few months, when the "last" bugs were ironed out, xTc_Rex.py became of vital importance for the outcome of this work.

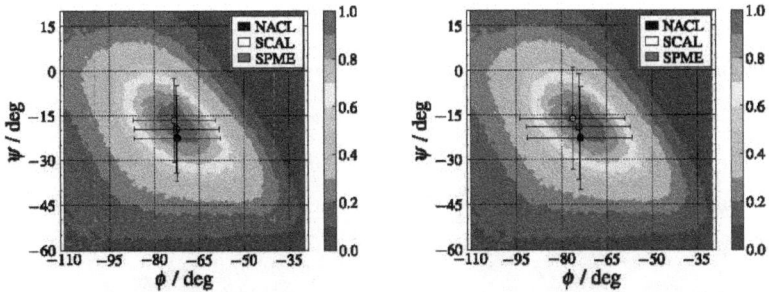

(a) *Ox. hIAPP Ramachandran Plot $60° \times 60°$ at* 350 K (b) *Ox. hIAPP Ramachandran Plot $80° \times 80°$ at* 350 K

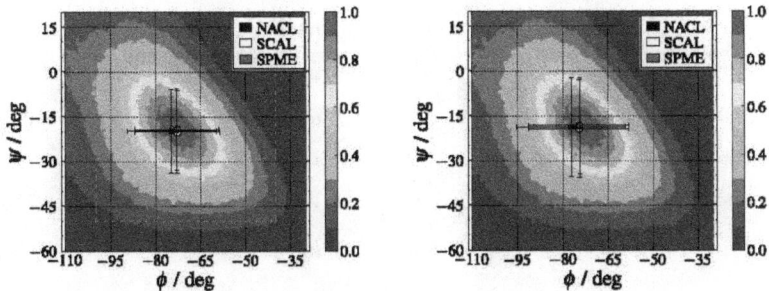

(c) *Red. hIAPP Ramachandran Plot $60° \times 60°$ at* 350 K (d) *Red. hIAPP Ramachandran Plot $80° \times 80°$ at* 350 K

Figure 2.4: *(a)–(b) Oxidized and (c)–(d) reduced hIAPP Ramachandran plots relative to the helical region enclosed by the dashed red line, $-100° \leq \phi \leq -40°$ and $10° \leq \psi \leq -50°$, for (a) and (c), and $-110° \leq \phi \leq -30°$ and $20° \leq \psi \leq -60°$, for (b) and (d) for NACL, SPME, and SCAL runs. The circles indicate the mean value of the ϕ and ψ angles for the data within the dashed red square, with the area of each circle proportional to the contribution of each trajectory to the data distribution. The error bars indicate the standard deviation of the mean.*

and the chirality of the disulfide bond dihedral angle, whereas the generic properties, e.g., R_g, r_{eted}, and the number of intrapeptide H-bonds and H-bond matrices, can be calculated for any monomeric polypeptide.

2.4.1 Ramachandran Angles

Besides using SEGNO[79] and DSSPcont[77,78] to assign secondary structure elements, regions of the Ramachandran plot can also be used to define secondary structure according to the ϕ and ψ backbone angles.[75] The residues are assigned a particular secondary structure if the Ramachandran angles are the following, as seen in Figure 2.7: $-100° \leq \phi \leq -40°$ and $10° \leq \psi \leq -50°$ correspond to helical conformations, $-175° \leq \phi \leq -115°$ and $125° \leq \psi \leq 185°$ correspond to β-strands, and $-115° \leq \phi \leq -55°$ and $105° \leq \psi \leq 165°$ correspond to poly(L-proline) I and II helices, where poly(Pro) I con-

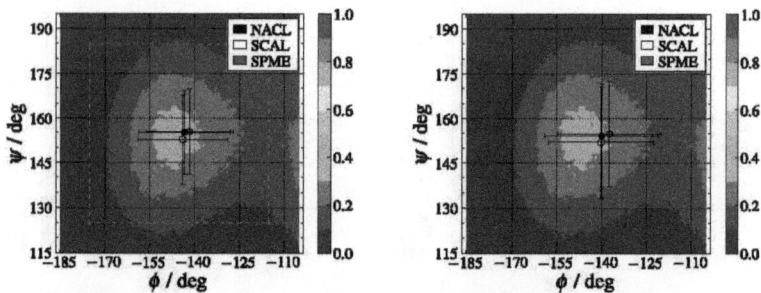

(a) *Ox. hIAPP Ramachandran Plot 60° × 60° at 350 K* (b) *Ox. hIAPP Ramachandran Plot 80° × 80° at 350 K*

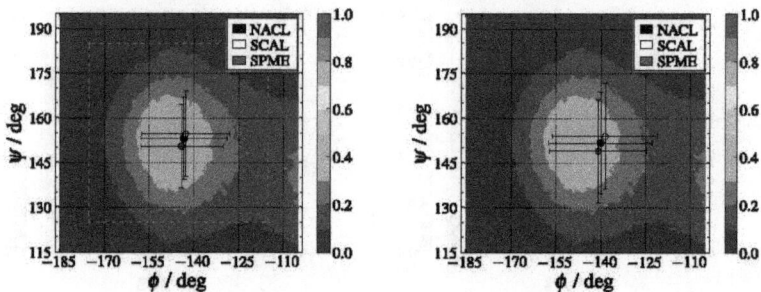

(c) *Red. hIAPP Ramachandran Plot 60° × 60° at 350 K* (d) *Red. hIAPP Ramachandran Plot 80° × 80° at 350 K*

Figure 2.5: *(a)–(b) Oxidized and (c)–(d) reduced hIAPP Ramachandran plots relative to the isolated β-strand region enclosed by the dashed red line, $-175° \leq \phi \leq -115°$ and $125° \leq \psi \leq 185°$, for (a) and (c), and $-185° \leq \phi \leq -105°$ and $115° \leq \psi \leq 195°$ for (b) and (d) for NACL, SPME, and SCAL runs. The circles indicate the mean value of the ϕ and ψ angles for the data within the dashed red square, with the area of each circle proportional to the contribution of each trajectory to the data distribution. The error bars indicate the standard deviation of the mean.*

sists of proline residues that contain only *cis* bonds, while poly(Pro) II contains only *trans* bonds. The theoretical values calculated for rigid spheres and van der Waals radii[74,75] show a nice correspondence with the chosen cutoffs for poly(Pro) and slightly less for β-sheets. The extended conformations are slightly shifted to more positive values for ψ, and more negative values for ϕ, with the parallel β-sheets not sampled, possibly due to the highly improbable conformation necessary for parallel alignment of two segments in such a short peptide. Moreover, the antiparallel β-sheet is seen to be more stable between the two secondary structures. The β-sheets, initially thought to be flat and planar, present a right-handed twist, hence yielding slightly more positive values for ϕ and ψ.[5] The helical conformations also deviate from the theoretical values and normally present ϕ equal to $-62°$ and ψ $-41°$. This deviation is due to the carbonyl groups pointing outward, away from the helix, and the H-bonds are not as straight in order to create a more favorable geometry that allows H-bond formation with water molecules or other H-bond

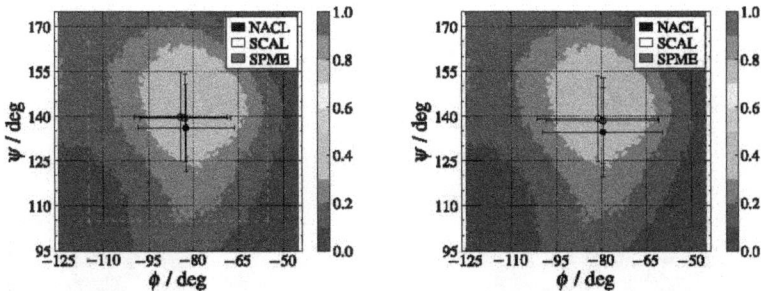

(a) *Ox. hIAPP Ramachandran Plot* $60° \times 60°$ *at* 350 K (b) *Ox. hIAPP Ramachandran Plot* $80° \times 80°$ *at* 350 K

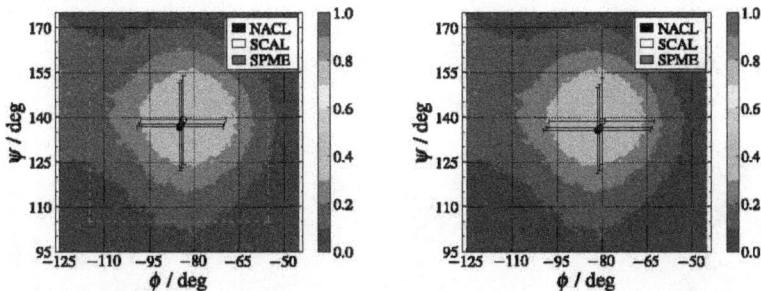

(c) *Red. hIAPP Ramachandran Plot* $60° \times 60°$ *at* 350 K (d) *Red. hIAPP Ramachandran Plot* $80° \times 80°$ *at* 350 K

Figure 2.6: *(a)–(b) Oxidized and (c)–(d) reduced hIAPP Ramachandran plots relative to the poly(Pro) II region enclosed by the dashed red line, $-115° \leq \phi \leq -55°$ and $105° \leq \psi \leq 165°$, for (a) and (c), and $-105° \leq \phi \leq -45°$ and $95° \leq \psi \leq 175°$, for (b) and (d) for NACL, SPME, and SCAL runs. The circles indicate the mean value of the ϕ and ψ angles for the data within the dashed red square, with the area of each circle proportional to the contribution of each trajectory to the data distribution. The error bars indicate the standard deviation of the mean.*

donors.[5] An additional note on H-bond formation in helices, certain residues, i.e., serine, threonine, aspartic acid, and asparagine, can interfere in α-helical H-bonding since the side chain can bend back and form bonds with the backbone atoms. In fact, these residues are often found in capping the α-helix, as there are four peptide oxygens or hydrogens that are not involved in helix bonding.[5] Moreover, it is not surprising that some of the conformations present ϕ and ψ angles in the restricted areas of the Ramachandran plots, because the bonds and angles are flexible and can stretch and bend to avoid clashing of spheres held together by rigid bonds, used to determine the permitted values in the Ramachandran plots. These classical values are depicted by green circles in Figure 2.7, namely antiparallel β-sheet $(-139°, 135°)$, parallel β-sheet $(-119°, 113°)$, α_R-helix $(-57°, -47°)$, 3_{10}-helix $(-49°, -26°)$, π-helix $(-57°, -70°)$, poly(Pro) I $(-83°, 158°)$, and poly(Pro) II $(-78°, 149°)$. Ramachandran plots for known and resolved proteins that present high α-helical content, Soluble Lytic Transglycosylase Slt70 (1QSA), and

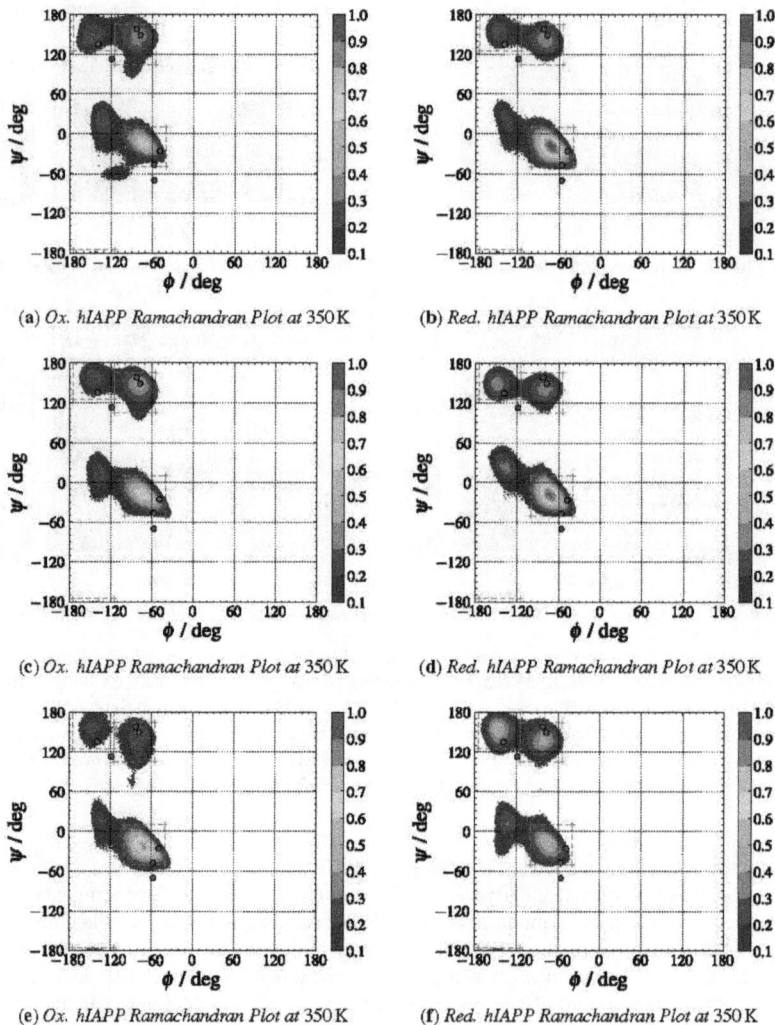

(a) *Ox. hIAPP Ramachandran Plot at* 350 K

(b) *Red. hIAPP Ramachandran Plot at* 350 K

(c) *Ox. hIAPP Ramachandran Plot at* 350 K

(d) *Red. hIAPP Ramachandran Plot at* 350 K

(e) *Ox. hIAPP Ramachandran Plot at* 350 K

(f) *Red. hIAPP Ramachandran Plot at* 350 K

Figure 2.7: *Ramachandran plots for the oxidized hIAPP (a) SCAL, (c) SPME, and (e) NACL runs, and for the reduced hIAPP runs (b) SCAL, (d) SPME, and (f) NACL. The green circles indicate the theoretical values calculated for rigid spheres and van der Waals radii.[74,75] The occurrences have been normalized and are relative to the highest peak found in (b).*

high β-sheet content, Pectin Lyase A (1IDK), show that the cutoffs chosen to encompass the helical portion of the Ramachandran plot are appropriate, with the most occurring angles being slightly shifted, as suggested by Creighton. On the other hand, the area of the Ramachandran plot relative to the β-sheets presents points that are not encompassed by the chosen cutoff, since the Pectin Lyase A presents parallel β-sheets, which are not present in hIAPP. Many data points are also present in the cutoff region assigned to poly(Pro), even when glycine and proline are not plotted, but this region is also populated by many other structures available at the PDB database, for example the PDB40 dataset,[81] as plotted in Molecular Modeling and Simulation by Schlick.[2]

By assigning the secondary structure through Ramachandran angles, there is no differentiation between 3_{10}, α, or π helices, as the H-bond pattern is not taken into account to determine which kind of helix the conformation assumes. Both SEGNO[79] and DSSP-cont[77,78] assign secondary structure elements following different algorithms and do not rely solely on Ramachandran angles. In fact, DSSP[77] relies on H-bonds, since H-bonds depend only on the cutoff in energy and not on positions of C_α or on ϕ and ψ backbone angles. Moreover, only a minimum number of consecutive residues with the same conformation are assigned to a secondary structural element. Even if the determination of the secondary structure determined by the selected areas of the Ramachandran plot may overestimate the helical content by including H-bonded turns, which otherwise would not be assigned as a helix, such criteria may be relevant for certain analyses, as residues that participate individually in the cooperative construction of a certain conformation are also included and is not limited to those that participate only in a complete conformation.

Table 2.1: *Secondary Structure Assignment*

Assignment	t / ns	KCNTATCATQRLANFLVHSSNNFGAILSSTNVGSNTY
DSSPcont	7.834	LLTTSLGGGGGGGGGHHHHTTTSLLSGGGTTTLLLLLL
Rama. Angles	7.834	LLHLPPHHHHHHHHHHHLHHLLLILHHLLLHLILLPLL
SEGNO	7.834	LLLLLLGGGGGGGGGGGGGGGGLLLLLLLLLLLLLLLL
DSSPcont	10.580	LLTTSLSSTTTSLHHHHTHHHHHHHTLLSSLSSSSLL
Rama. Angles	10.580	LLHLIPLLHHLIPHHHHHHHHHLHLLLLLHIIHLHIPL
SEGNO	10.580	LLLLLLLLLLLLMMMMMMMMMMMLLLLLLLLLLLLLLL
DSSPcont	1.006	LLTTSLLSGGGGGTTSSIIIIIIIITTGGGGLLLLLL
Rama. Angles	1.006	LLHLLPILHHHHLLLLPLHLLHLLHHLHHHHPLIPLL
SEGNO	1.006	LLLLIIILGGGGLLLLLLLLLLLLLLLLGGGGLLLLLL

As previously mentioned, these classical angles are defined in solid structures and do not correspond exactly to values in solution, hence slightly different values were taken from those used in SEGNO; in fact, the $60° \times 60°$ regions of ϕ and ψ that encompassed the maximum of the Ramachandran plots in Figure 2.7 were chosen. In Table 2.1, three conformations at arbitrary time steps have been selected to illustrate the secondary structure assignment of helices by the Ramachandran angles (in red), SEGNO (in cyan), and DSSPcont (in green). The helices (H) defined by the Ramachandran plot correspond in many cases with both SEGNO and DSSPcont, especially for 3 and 4-turn helices, G and H respectively, even though there is a slight overestimation of helices compared to DSSPcont, since turns (T) are hydrogen bonded turns that do not have sufficient continuous residues in the same secondary structure or may be in a region of overlap between two sequences of other helices.[77] A geometrical structure assigned by DSSPcont, which can be assigned as a helix by the Ramachandran angles, is the bend (S), a property not determined by H-bonds, but rather simply by the angle between the first, the third, and the fifth C_α of five consecutive residues and is assigned as *bend* if the angle is less than 70°. Residues can be assigned as helical (H) by the Ramachandran angles even in the case of bends assigned by DSSPcont, resulting therefore in an overestimation of the helical content. It is important to notice, however, that coils or loops (L), i.e., secondary structure that cannot be assigned to any regular structure through any of the definitions found in DSSPcont, have not been assigned as H by the Ramachandran angle selection. There is also a good correspondence between Ramachandran angles that define helices (H) and mixed helices (M), assigned by SEGNO, showing albeit a slight overestimation by the Ramachandran angle assignment, as can be seen in the second set of data. Unfortunately, the third data set shows that the sampling of the extremely rare 5-turn helix is not reproduced as clearly; in fact, the sequence of 5-turn helices detected by DSSPcont (I) is not detected by SEGNO either. The π-helix is rare because the backbone is so loose that it creates a cavity along its axis.[2]

2.4.2 Hydrogen Bond Definitions

Not only are hydrogen bonds an important structural property, but they also form patterns that correlate highly to secondary structure. Such patterns can be either limited to the backbone-backbone interactions or extended to backbone interactions with the side chains. Considering that normally only heavy atoms are detected when resolving X-ray structures, H-bonds can be expressed by the distance between heavy atoms in H-bonds, i.e., approximately $3\,\text{Å}$.[5] The typical H-bond distances, as can be found in Table 2.2 of Biochemistry by Stryer,[1] are as follows:

Table 2.2: *Hydrogen Bonds in Biological Systems*

bond	length / Å	bond	length / Å
O−H···O	2.70	N−H···O	3.04
O−H···O⁻	2.63	N⁺−H···O	2.93
O−H···N	2.88	N−H···N	3.10

The main component of H-bonds is the electrostatic interaction between the positive partial charge on H and the negative partial charge on the donor (D) and acceptor (A) atoms, as such $D^{\delta-}-H^{\delta+}\cdots A^{\delta-}$.[5] The scaling of the charges to neutrality on a positive peptide make the negative charges more negative and the positive charges less positive, so it is possible that the D−A is slightly longer than what is found in Table 2.2, because of to the larger repulsion between D and A and less screening of H between them. Comparing the backbone-backbone H-bond content obtained with the less stringent criteria described in the following paragraph, there is little or no correlation between the H-bonds of the parallel independent runs illustrated in Section 2.3 ($|r_{xy}| \leq 0.2$). The Bland-Altman plots[72] show comparable H-bonds ($|\Delta n_{HB}| \approx 2$, with $n_{HB(D)} > n_{HB(G)}$

and $n_{HB(D)} \leq n_{HB(P)}{}^*$) with no apparent bias (not shown).

Backbone-backbone H-bonds were identified by the following two cutoff criteria: distances between the donor (D) and the acceptor (A) atoms less than 3.5 Å and donor-hydrogen-acceptor atom angles, DHA, greater or equal to 130°. Oxygen and nitrogen atoms were classified as donors if bound to hydrogen and as acceptors if a lone pair was present. Many H-bonds in proteins involve the carbonyl C=O and the amide N−H groups of the peptide bond, and because of the lone pair in the carbonyl group, the H-bonds have a preferred geometry with H on the plane defined by $R_2C=O$, normally within 0°–7°, and deviating from the linearity of C=O···H by 50°–60°.[5] Therefore, a more stringent definition was used to determine backbone-backbone H-bonds, similar to the one used by Kuster et al.[82] The carbonyl carbon of the backbone is also included in the acceptance criteria of the DHA angle, i.e., both the C=O···H and O···H−N angles in C=O···H−N must be greater than 90°, and one of the two must be greater or equal to 130°. The H-bond residue-residue matrices in Figures 5.18 and 5.19, which can be related to secondary structure, were assigned with this criteria. Such plot is defined by registering the occurrence of backbone-backbone H-bond between two residues, which are aligned along the two axes according to the peptide sequence from first to last. If the data points lie along the main diagonal, a helix may be identified. The helix, if present, may then be classified according to the number of residues between a pair, i.e., four residues define an α-helix. If the data points lie perpendicular to it, such behavior is characteristic of another secondary structure, like a β-sheet.

2.4.3 Statistical Properties

Important information on the conformational behavior of a random polypeptide can be obtained by statistically averaging all the visited conformations; two such properties are end-to-end distance, r_{eted}, and radius of gyration, R_g. The former is the root-mean-square, RMS, of the distance between the first and last atoms of the chain, while the latter is defined as the RMS of the distance of each atom from the center of mass of the polypeptide.

As a first approximation, statistical properties of a random polypeptide chain, such as end-to-end distance, r_{eted}, or the radius of gyration, R_g, can be calculated for an **unperturbed random coil**, using a **random-flight chain model**, in which each amino acid residue is not influenced by the remaining residues of the polypeptide nor is the excluded volume effect taken into account. A simplified model can thus be defined: atoms are not included, chains have no volume, all bond angles are equally probable, as are the rotations around bonds, and bond lengths, l, are set to 0.38 nm, which is the distance between adjacent C_α in a planar *trans* conformation. The end-to-end distance (Eq. 2.8) and radius of gyration (Eq. 2.9) for such a model of n bonds are as follows:

$$\langle r_{eted}^2 \rangle_0^{1/2} = n^{1/2} l, \tag{2.8}$$

$$\langle R_g^2 \rangle_0 = \frac{nl^2}{6}\frac{n+2}{n+1}, \tag{2.9}$$

*The number of H-bonds is labeled as for the electrostatic potential: scaled charges trajectory $n_{HB(D)}$, opposite partial charges distributed on the SPME grid $n_{HB(G)}$, and neutralizing charges given by a NaCl solution at physiological ionic strength of 150 mM $n_{HB(P)}$.

where the angle brackets indicate the statistical mechanical average over all conformations and the subscript zero refers to the unperturbed state.[5] Moreover, close examination of Eqs. (2.8) and (2.9), shows a relationship between these two statistical properties, which for a short chain is as follows:

$$\frac{\langle R_g \rangle_0^2}{\langle r_{eted}^2 \rangle_0} = \frac{1}{6}\frac{n+2}{n+1}. \tag{2.10}$$

The statistical properties described by Eqs. (2.8)–(2.10), calculated with l set to 0.38 nm and n to 36, are listed in Table 2.3, where the values on the left are squared, while on the right they are to the first power.

Table 2.3: *Random-flight Chain Statistical Properties (*$n = 36$, $l = 0.38$ nm*)*

Statistical Property	Calculated Value	Statistical Property	Calculated Value
$\langle r_{eted}^2 \rangle_0$	5.2 nm^2	$\langle r_{eted}^2 \rangle_0^{1/2}$	2.3 nm
$\langle R_g \rangle_0^2$	0.89 nm^2	$\langle R_g \rangle_0$	0.94 nm
$\langle R_g \rangle_0^2 / \langle r_{eted}^2 \rangle_0$	0.17	$\langle R_g \rangle_0 / \langle r_{eted}^2 \rangle_0^{1/2}$	0.41

The radius of gyration increases proportionally to the chain length with a clear influence of the peptide conformation on how it increases. As seen in Figure 5.3 of Proteins: Structures and Molecular Properties, by Creighton (1993), a polypeptide chain of 37 amino acid residues presents smaller R_g when in spherical conformation, while it is larger when in α-helical or random coil conformation. Hence, some information on the conformation of the polypeptide can be given by the average value of the radius of gyration. Moreover, inherent properties of amino acid residues and limited flexibility of polypeptide chains determine a **characteristic ratio**, C_n, that exists between the measured end-to-end distance and the one calculated from the random-flight chain as seen in Eq. (2.11):[5]

$$C_n = \frac{\langle r_{eted}^2 \rangle_0}{nl^2}, \tag{2.11}$$

which for infinite chains, C_∞ is 9.0. This is true if no glycine or proline residues are present: Proline, in the *cis* form, induces a change in direction of the chain and can lower C_∞; interspersed glycine, lacking chiral side chains, can also make the chain more flexible, lowering the value of C_∞.

The distribution of the possible values of r_{eted} assumed by the random-flight can be described by a radial distribution, $W(r)dr$, which is the probability of the two ends to be between a distance r and $r + dr$, as can be seen in the following equation:[5]

$$W(r)dr = \left(\frac{\beta}{\sqrt{\pi}}\right)^3 e^{-\beta^2 r^2} 4\pi r^2 dr, \tag{2.12}$$

where $r^2 = x^2 + y^2 + z^2$ and $\beta = \left(\frac{3}{2nl^2}\right)^{\frac{1}{2}}$. Such kind of distribution is asymmetrical and the RMS value of r does not coincide with the maximum, as can be seen in Figure 2.8. A finite chain will not follow this distribution, since there is a non-zero probability of the r_{eted} to be larger that the length of the chain.

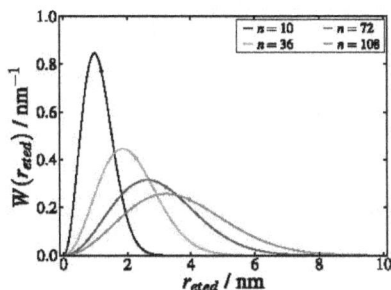

Figure 2.8: *Radial distribution, Eq. (2.12), for various random-flight chains of arbitrary lengths, with $n = 36$ in cyan. As the chain length increases, so does the breadth of the distribution. The RMS values of the distance between two atoms, $\langle r^2_{eted}\rangle_0^{1/2} = n^{1/2}l$, are represented by a point on each distribution. Owing to the asymmetry of this radial distribution, the RMS distance is not the most probable value.*

2.4.3.1 Water box and Maximum Distance between Heavy Atoms

The upper limit of r_{eted} is given by the maximum distance between heavy atoms, as seen in Figure 2.10 on page 37, so L_{max} can determine the maximum extension of the polypeptide and is useful to monitor whether the box size of the simulation is sufficiently large to avoid short-range interactions between the images; in fact, only the nearest image in periodic boundary conditions is considered when calculating interactions between atoms in systems in which the minimum image convention is applied. In other words, the implemented algorithm of periodic boundry conditions, combined with the minimum image convention, does not distinguish between the interpeptide interactions of two images and the intrapeptide interactions of a monomer. The short-range nonbonded interactions are calculated only for atoms within a cutoff of 0.9 nm, therefore it is important that two images do not interact with each other by forming interpeptide interactions between the termini, i.e., the C-terminus interacting with the N-terminus of its virtual image would erroneously simulate "two" peptides in solution, rather than a monomer, creating thus a periodic artifact of an infinite chain of interacting polypeptides.

A $216\,nm^3$ water box containing 7161 SPC/E[60] water molecules was prepared to solvate four different conformations obtained from the run in vacuum for both moieties, i.e., from RUN1 to RUN4. The box of water molecules included in the GROMACS library consists of 216 SPC water molecules equilibrated at 300 K with a box edge of 1.862 06 nm prepared by van Gusteren.[*] Since the simulation run is set at 350 K, the water box must be heated after being minimized and equilibrated (L-BFGS and CG minimizations, *NPT* equilibration and production run). The starting density at 300 K was 991.8 kg m^{-3}, which after 1 ns at 350 K resulted in (965 ± 3) kg m^{-3}. Such value seems a bit strange; even if SPC/E water presents a density lower than experimental values,[83] that seems a little too low. Since the water box was heated at constant pressure, it is possible that the box size expanded too much lowering the density excessively. In fact, after solvating the peptide with such water box and pre-equilibrating the system at constant volume at 350 K, the average density for a 20 ns trajectory increases to (970.89 ± 0.03) kg m^{-3}, which is closer to that of pure water calculated[†] at 350 K, 973.68 kg m^{-3}. A temperature-dependence

---

[*]The only details found in the GROMACS library are "216H2O, WATJP01, SPC216, SPC-MODEL, 300K, BOX(M)=1.86206NM, WFVG, MAR. 1984."

[†]The water density used as comparison to the SPC/E model was obtained with the water density calculator found at the following website: http://www.csgnetwork.com/h2odenscalc.

plot of the average values of water density can be seen in Figure 2.9, where the calculated values are indicated with full circles and the corresponding polynomial fitted curves indicated by dashed lines, where the density maxima are indicated by diamonds. Many water models fail to reproduce the water density maximum of $0.99995\,\mathrm{g\,cm^{-3}}$ at $4\,°\mathrm{C}$;[84] in fact, SPC/E presents a maximum at $235.15\,\mathrm{K}$,[85] which is also extrapolated to $233.75\,\mathrm{K}$ with the polynomial fitted curve. The polynomial fit of the calculated pure water points also reproduces the density maximum pretty well with $998.41\,\mathrm{kg\,m^{-3}}$ at $280.50\,\mathrm{K}$.

Figure 2.9: *Water density for pure water and a hIAPP SPC/E 3.0% w/w solution for temperatures between 250 K and 450 K, with the polynomial fit $y = ax^3 + bx^2 + cx + d$ in dashed line. The maximum of the water density at 277.15 K obtained from the polynomial fit curve is 999.347 kg m^{-3} (indicated with a plus sign), compared to the experimental values of 999.9750 kg m^{-3} (SMOW)[84] and 999.95 kg m^{-3} ($H_2O_{(l)}$).[84] The maxima of the polynomial fits are indicated by diamonds and are 1020.7 kg m^{-3} and 998.41 kg m^{-3} at 233.75 K and 280.50 K, respectively.*

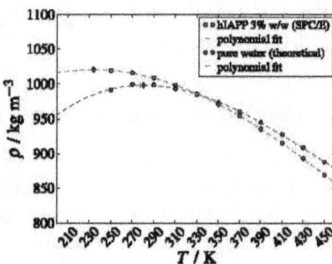

The system box size was set to 6 nm for runs RUN1 to RUN4 and 7 nm for RUN0, since the initial α-helical conformation was longer and expected to explore more extended conformations than the more compact ones obtained in vacuo. If L_{max} were to reach values comparable to the box size, one could expect eventual short-range interactions between images of the peptide, but as can be seen in Figure 2.10a, both the mean value of L_{max} and the error bars, indicating three standard deviations, do not cross the dashed cyan line representing a distance of 0.9 nm from the average values of the side of the box. Thus, it is possible to say that, within the confidence limit of 99.7 %,[86] the conformations do not interact with their images. Hence, 7 nm seems to be the correct choice for the box size, confirmed also by the trajectories of the capped peptide, which explore larger values of r_{eted} more frequently, yet without crossing the cutoff limit throughout 50 ns of the MD simulation, as can be seen in Figure 2.10b.

2.4.4 Data Crunching

The output of the software used was subsequently parsed and processed through other ad hoc python scripts, with the use of scientific modules and packages, e.g., SciPy and NumPy, and plotted with MatPlotLib. The error estimation was carried out by applying g_analyze.[80]

2.4.4.1 Error Estimate

Estimating the error of properties calculated in MD simulations is not trivial, especially because an enormous amount of data can be generated; moreover, it most likely is highly

---

html.

correlated. A first approximation is finding the standard deviation, which is the square root of the variance:[45]

$$\sigma_{\langle A \rangle} = \frac{\sigma_A}{\sqrt{M}} = \sqrt{\frac{\sum_{i=1}^{M}(A(i) - \langle A \rangle)^2}{M}}, \tag{2.13}$$

where the standard deviation of the average value of property A is obtained by dividing by the square root of the M data values, which are not the number of data points. In fact, to be able to use this formula to estimate the standard deviation of the mean and thus the error, the data used must be independent. The data in MD simulations are highly correlated, and to properly calculate the mean, the data point needs to be collected after the correlation time, i.e., after it has lost "memory" of its previous state. In other words, it is necessary to wait for the relaxation time or correlation time to pass before using another data point.

One way to sample the data is to create blocks of data of time, τ_B, large enough for them to be statistically uncorrelated. The number of these blocks, M, will determine how many data points can actually contribute to calculating the "true" standard deviation, from which it is possible to calculate the "true" error. The number of correlated points can be given by the *statistical inefficiency*, s, determined by the number of data points that are correlated, e.g., if $s = 22$ then only one configuration every twenty-two of the collected data brings new information to the measured property.[87] If the statistical inefficiency is calculated, then the error can also be calculated by means of the following equation:[45]

$$\delta_{\langle A \rangle} \approx \sigma \sqrt{\frac{s}{M}}, \tag{2.14}$$

where M is actually the number of data points.

Calculating the statistical inefficiency for each property can be a bit tedious, because one needs to know the correlation time with which the blocks are made. Luckily, g_analyze[80] is a complete tool that calculates both the error and the autocorrelation times. The variance of the average values $\langle B_i \rangle$ for each of the m blocks[87]

$$\sigma^2 = \frac{\sum_{i=1}^{m}(\langle B_i \rangle - \langle B \rangle)^2}{m} \tag{2.15}$$

is used to calculate the error on the total average, as seen in the following equation:[69]

$$error = \sqrt{\frac{\sum_{i=1}^{m}(\langle B_i \rangle - \langle B \rangle)^2}{m(m-1)}}. \tag{2.16}$$

If the autocorrelation can be expressed as a sum of two exponentials,* the analytical curve for the block average is[69]

$$f(t) = \sigma \sqrt{\frac{2}{T} \left\{ a\tau_1 \left[\left(e^{\frac{-t}{\tau_1}} - 1 \right) \frac{\tau_1}{t} + 1 \right] + (1-a)\tau_2 \left[\left(e^{\frac{-t}{\tau_2}} - 1 \right) \frac{\tau_2}{t} + 1 \right] \right\}}, \tag{2.17}$$

---

*This is derived from studies involving Principle Component Analysis on a system in which the principle components showed both rapid and slow fluctuations, the former of the order of tens of picoseconds and the latter of hundreds of picoseconds. Thus, the autocorrelation functions of the principle components can be fitted by two exponentials, one with fast and the other with slow correlation times. In other words, there are two stochastic processes with white noise and time independent friction constants.[80]

where T is the total time, and a, τ_1, and τ_2 are the parameters obtained by fitting the square of Eq. (2.17) to the square of Eq. (2.16). If the actual block average is very close to its analytical curve, i.e., Eq. (2.17), the error can be given by the following equation:[69]

$$\delta_{\langle B \rangle} = \sigma_{\langle B \rangle} \sqrt{\frac{2}{T} [a\tau_1 + (1-a)\tau_2]}. \tag{2.18}$$

Current MD simulations are limited to hundreds of nanoseconds, therefore a complete sampling is not possible as proteins in solution undergo fluctuations that can range from femtoseconds to seconds, depending on the measured property. Fluctuations around one conformation are easily measured, but to actually see if the system is doing something more is difficult and needs plenty of time. A few examples of the errors calculated on a few properties obtained from the 65 nm reduced hIAPP RUN0 can be seen in Table 2.4.

Table 2.4: *Error Estimate Calculated by* g_analyze

	$\sigma_{\langle B \rangle}$	$\delta_{\langle B \rangle}$	a	τ_1 / ps	τ_2 / ns	T / ns	δ/σ
$n_{HB(tot)}$	3.30	0.682	0.678	20.89	4.27	65.0	0.207
$n_{HB(bb)}$	3.30	3.73	0.361	75.06	65.0	65.0	1.13
SASA / nm^2	1.15	1.31	0.355	97.79	65.0	65.0	1.14
r_{eted} / nm	1.16	0.583	1.0	8213	0.0	65.0	0.503

When the fitting presents no negative parameters, the upper limit of the error is the standard deviation, and the more the data are uncorrelated, the greater m in Eq. (2.16) is, and the smaller τ_2 in Eq. (2.17) becomes. In fact, the first row of Table 2.4 shows that the number of total hydrogen bonds is well sampled and the δ/σ ratio, which is the term enclosed by the square root in Eq. (2.18), is smaller than one. Although the fitting can lead to negative parameters, especially for highly correlated data or insufficient sampling, Eq. (2.18) can nevertheless be approximated to

$$\delta_{\langle B \rangle} \approx \sigma_{\langle B \rangle} \sqrt{2(1-a)} \tag{2.19}$$

when τ_2 is either too long for the sampled data or negative, as seen in rows two and three in Table 2.4 relative to backbone-backbone hydrogen bonds and solvent accessible surface. In either case, τ_2 is set to the simulation time. When a is negative, it is set to one and Eq. (2.18) can be approximated to

$$\delta_{\langle B \rangle} \approx \sigma_{\langle B \rangle} \sqrt{\frac{2}{T} \tau_1} \tag{2.20}$$

which is still dependent on the fast correlation time, τ_1. In the best case scenario, the data are sampled sufficiently well to allow a reasonable error, as seen in the last row in Table 2.4 relative to r_{eted}, where τ_1 is still shorter than the total simulation time, T. When the data are seriously suspicious and the fast correlation time is of various orders of magnitude larger than the simulation time, the error can be incredibly large, as can be seen in Figure 3.5, which will be discussed in Section 3.2.

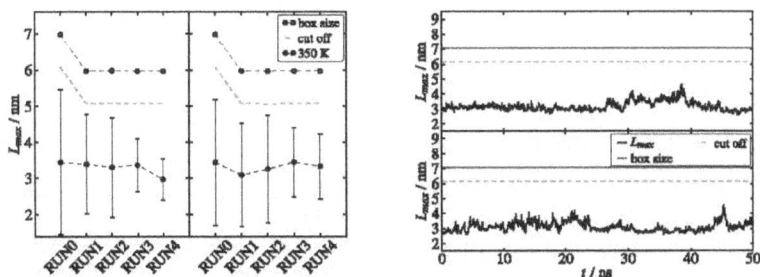

(a) *Maximum Distance between Heavy Atoms at* 350 K(b) *Maximum Distance between Heavy Atoms at* 350 K

Figure 2.10: *(a) The mean maximum distance between heavy atoms, L_{max} (black), at 350 K, compared to the box size of the simulation (blue) and the 0.9 nm short-range interaction cutoff in relation to the box size (cyan) for each of the uncapped runs. The error bars show the confidence limit of 99.7 %. The left panel shows the oxidized hIAPP, while reduced hIAPP is in the right panel. (b) Time-dependence of L_{max} (black) for capped hIAPP at 350 K compared to the box size (blue) and the SPME cutoff in relation to the box size (cyan). The top panel shows the oxidized hIAPP, while reduced hIAPP is in the bottom panel.*

2.4.4.2 Savitzky-Golay Smoothing Filter

The smoothing on all the time-dependent data is performed with an implementation of the Savitzky-Golay Smoothing Filter[88] python script available from http://public.procoders.net/sg_filter. The data are smoothed by an implementation of local polynomial regression of degree k on a series of at least $k+1$ data points. The data are reduced only to remove the fluctuations in order to aid visualization in the time-dependent plots, but have been processed in their integrity when, for example, averaged over time or binned in histograms/distributions.

2.4.4.3 Scott's Choice

The best method for binning the time-dependent data was Scott's Choice,[89] which defines the bin width of the histogram, h, in function of the sample standard deviation, σ, and the number of total observations, n, as follows:

$$h = \frac{3.5\sigma}{n^{1/3}}. \tag{2.21}$$

2.4.4.4 Rounding Data — Taylor

It is interesting to note that the data have been rounded by keeping an extra significant figure when the numbers are small, i.e., one or two, and the first figure of the error is one, as suggested by Taylor.[70] An example can be seen when calculating the end-to-end distance, as seen in Table 3.2 on page 46. The error is large in these data as a result of the difficulty in calculating r_{eted}, but it would be even more so to round (1.6 ± 1.2) nm and (1.4 ± 1.2) nm, to (2 ± 1) nm and (1 ± 1) nm, respectively. Therefore, an extra figure was kept when the first error was one when presenting the data.

2.5 Hydration Water

2.5.1 System Description

Oxidized and reduced human IAPP were simulated in liquid water at 11 temperatures between 250 K and 450 K. After a 50 ns pre-equilibration at each temperature, the subsequent runs for 400 ns and 200 ns were used for the analysis of the system properties at lower and higher temperatures, respectively. The data sampled at lower temperatures, i.e., 250 K to 350 K, were obtained by concatenating runs, i.e., ALL3, as described in Section 3.2, while at higher temperatures, i.e., 370 K to 450 K, a single run was sufficient, i.e., ALPH run as described in Section 3.2. Additionally, two conformations of reduced hIAPP, i.e., CL02 and CL03 seen in Section 3.2, exhibiting essentially different structural properties were used for simulations at 250 K. At such low temperature, the peptide conformation does not change noticeably during simulation runs of 50 ns allowing a comparison of the properties of hydration water at the surface of a peptide exhibiting different conformations. Oxidized and reduced rodent IAPP were simulated at 310 K and 330 K, i.e., in the temperature region where the strongest changes of the hydration water network are observed. After a pre-equilibration of at least 50 ns, the subsequent 500 ns were used for the data collection, as described for hIAPP in Section 2.2.3.

2.5.2 Water Shell Analysis Software

The water shell program was developed by Brovchenko and Oleinikova to analyze the hydration water around peptides. The connectivity of water-water H-bonds within the hydration shells of the IAPP variants was studied similarly to previous studies [40,90] by the analysis of the various clustering properties of hydration water. If the shortest distance between the water oxygen atom and the heavy atoms of the peptide does not exceed 0.45 nm, water molecules are assigned to the hydration shell. Two water molecules are considered as H-bonded when the distance between their oxygens does not exceed 0.335 nm and their pair interaction energy is below -2.7 kcal mol^{-1}. The connectivity of H-bonded network of hydration water at various temperatures is characterized by the occurrence probability n_S of water clusters consisting of S molecules and number n_H^{ww} of water-water H-bonds that a water molecule forms with its neighbors. Each configuration is examined in order to distinguish the largest cluster of hydration water of size S_{max}. The distribution n_S, calculated with the largest water cluster excluded, is used to determine the mean cluster size S_{mean}. The spanning probability, SP, is defined as the probability the largest water cluster has of including most of the N_w molecules in the hydration shell and is calculated based on the probability distribution of S_{max}/N_w. Additionally, the largest cluster of hydration water is characterized by the fractal dimension, d_f, and by the distance between its center of mass and the center of mass of a peptide, H_{max}.

The conformation of the peptide is characterized by the properties calculated by the program described in Section 2.4, namely the radius of gyration, R_g, the solvent accessible surface area, SASA, and the number of intrapeptide H-bonds, n_H^{pp}. The width of the probability distribution, $P(A)$, is calculated as $\Delta A = \sqrt{\langle (A - A^{av})^2 \rangle / N_w}$, where A stands for S_{max}/N_w, n_H^{pp}, or H_{max}.

Chapter 3

Preparation of the Initial Conformations

3.1 Random Conformations from Vacuum

Full-length hIAPP is a relatively large polypeptide to simulate in MD simulations with many interactions amongst the amino acid residues that seem to create problems in determining the conformation of IAPP even experimentally. As preliminary analysis to determine possible starting conformations of both polypeptide moieties, the radius of gyration and end-to-end distance between C_α were studied on a first set comprising five independent initial conformations, denominated RUN0, RUN1, RUN2, RUN3, and RUN4, with RUN0 being α-helical and RUN1 through RUN4 selected from families of structures using the GROMOS clustering algorithm[91] with an all-atom 0.8 nm RMSD cutoff* on two concatenated 1 ns runs at 1000 K in vacuo obtained in Section 2.2.1 and subsequently solvated in SPC/E water,[60] as described in Section 2.2.2. The initial conformations of these two runs were a fully extended peptide and an α-helical conformation.[†] A similar approach, although in preparation for a Replica Exchange Molecular Dynamics simulation, was used by Garcia to characterize non α-helical conformations in Ala peptides.[57]

The data of these isobaric-isothermal ensemble simulation runs were collected at 1 bar and 350 K, initially 15 nm for all five conformations, then three of them, from RUN0 to RUN2, were continued for additional 50 ns. Comparing the data of these relatively short runs to the theoretical values, obtained by using a **random-flight chain** as can be seen in Section 2.4.3 on page 31, it is possible to get a first estimate of the rigidity of the polypeptide and the intrapeptide interactions.

---

*The total number of clustered conformations was chosen in order to have a reasonable number of groups to choose from; with such cutoff, 11 clusters were generated. For the cystine moiety a structure from the first cluster was taken, characterized by a right-handed dihedral angle in the disulfide bridge ($\tau(C2_\beta S2_\gamma S7_\gamma C7_\beta)$), referred to as $\tau(S-S)$, and the other structures were taken from the third, fifth, and sixth clusters, representing the 73.95% of all visited conformations. As for the cysteine moiety, two conformations were taken from the first cluster, and the other two were taken from the second and third, representing 82.13% of the visited conformations.

[†]As a reference, the original α-helical structure was also solvated since simulation runs in vacuo might get "trapped" in certain conformations as a result of strong charge-charge interactions. An extended conformation was not run in water, considering the results obtained in vacuo do not depend strongly on the initial conformation, if not a possible effect on the chirality of the dihedral angle of the disulfide bridge $\tau(S-S)$. Both the right-handed and left-handed chiralities were simulated, as seen in Section 3.2 on page 50.

3.1.1 Data Analysis

3.1.1.1 Initial Modeled α-Helix Conformation

To investigate the statistical properties of hIAPP with and without the natural disulfide bridge between C2 and C7, five independent initial conformations were prepared for each moiety to explore conformational properties. Not only is this necessary to see the effect of the starting conformation on the outcome of the MD simulations, but it is also interesting to investigate the unfolding of this arbitrary α-helical conformation, because full or partial unfolding, rather than misfolding, seems to be a key step in amyloidogenic diseases.[19] Obviously, this state is taken only as a reference to an ordered state capable of "unfolding," since there is no known native structure of IAPP. Moreover, it may give some insight on the effect of the disulfide bridge on conformational properties and secondary structure of IAPP.

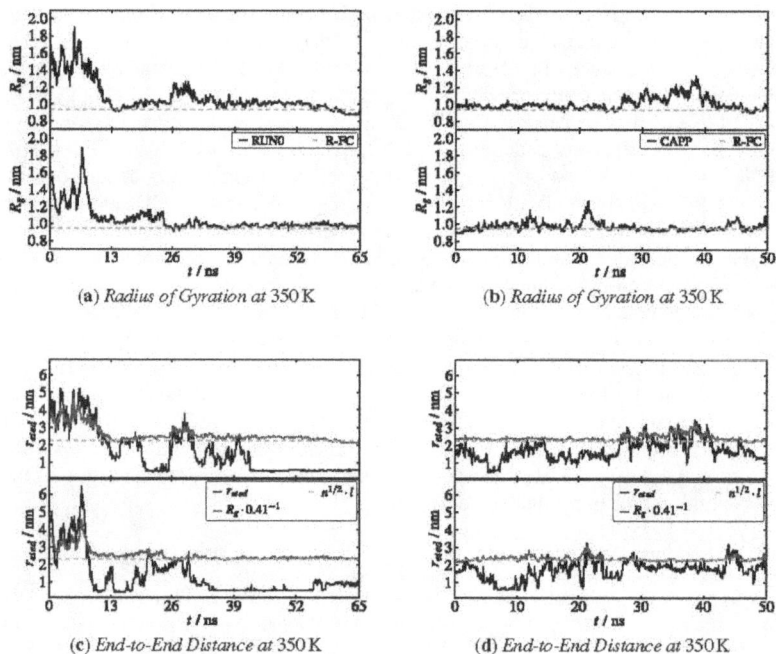

(a) *Radius of Gyration at* 350 K

(b) *Radius of Gyration at* 350 K

(c) *End-to-End Distance at* 350 K

(d) *End-to-End Distance at* 350 K

Figure 3.1: *The time-dependent data relative to oxidized hIAPP are found in the top panels of each figure, while the reduced hIAPP data are found in the bottom panels. (a) and (c) are relative to uncapped hIAPP, which was initially modeled as an α-helix, RUN0. (b) and (d) are relative to a completely random, but capped peptide. The blue plot in (c) and (d) is the theoretical value of r_{eted}, obtained from Eq. (2.10) in Table 2.3 and the measured R_g values. The dashed cyan line indicates the values calculated for the random-flight chain Eqs. (2.8) and (2.9) seen in Table 2.3.*

Fast unfolding and folding proteins may unfold on a 1–10 ns timescale at 100 °C,[92,93] so the dissolution of an initial α-helical conformation can be estimated to at least 10 ns at 350 K. In fact, using DSSPcont[77,78] to measure the helical content of the chosen conformers, the cysteine moiety, which was initially modeled in α-helical conformation, dissolves in 10.814 ns, while the cystine moiety, possibly in virtue of the presence of the natural disulfide bond, dissolves the initial helical content after 5.304 ns, far before reaching similar equilibrated values of R_g measured for both moieties, hence the cystine moiety assumes numerous other conformations before reaching a compact state. In fact, from the time-dependent data in Figure 3.1, it can be seen that the two moieties follow a different path in the dissolution of the initial modeled α-helix, labeled RUN0. The time-dependence of R_g of the oxidized moiety for RUN0 can be seen in black in the top panel of Figure 3.1a, and it is clear that R_g at $\approx$5 ns is still far from reaching the theoretical value calculated from the random-flight chain model seen in Table 2.3. In the top panel of Figure 3.1c, also relative to the oxidized moiety, r_{eted} of RUN0 fluctuates between 3 nm and 5 nm, indicating that it is exploring many different states while the helical content is diminishing. On the other hand, the reduced moiety reaches a compact state with a short r_{eted} prior to having dissolved the initial α-helical content, as seen at $\approx$10 ns in bottom panels of Figure 3.1a and Figure 3.1c, thus relatively independent from the helical content, i.e., the peptide already reached a compact state with short end-to-end distance before having completely dissolved the initial helical content. As seen in Figure 3.1a, especially in the second half of the runs, the data relative to R_g seems to fluctuate nicely around the theoretical value for both moieties as expected for an equilibrated system, while the data for r_{eted} seems to deviate more from its theoretical value. Considering the relation between r_{eted} and R_g expressed by Eq. (2.10) on page 32, it is possible to obtain an estimate of r_{eted} through the calculated R_g, as seen in blue in Figure 3.1c. These data seem to fit nicely onto the theoretical value of r_{eted} for the random-flight chain (cyan), which confirms the quality of the R_g data. The end-to-end distance is a difficult property to equilibrate, especially because the deviation from the theoretical value is highly dependent on the presence of glycine, present in positions 24 and 33 in the C-terminal half, possibly allowing the peptide to bring the two ends together more easily. As seen in the top panel of Figure 3.1c, the oxidized moiety seems more flexible and free to explore more values of r_{eted} than the reduced counterpart, as seen in the lower panel. One possible explanation of the relatively low sampling of r_{eted} is that the peptide was left uncapped for this run. Owing to the strong solvation properties of water, charges tend to be solvated by the surrounding water molecules, so the charged ends of this peptide could also be solvated by water. Although, a strong influence of conformational entropy will bring a random-flight chain to compact conformations, rather than extended ones. Termini in polypeptides when found in random coil state tend to be the most distant in random coil states; moreover, they are likely to be on the surface of the protein, accessible to water, and often flexible. On the other hand, when found in folded proteins the termini are frequently found closer together. In other words, the termini are more likely to be found closer together in natively folded peptides, rather than in randomly generated structures.[94] Therefore, it is possible that, even though the initial α-helical content dissolved at $\approx$10 ns, some of the structural characteristics of helicity are retained in RUN0, making it more rigid in the central region of the peptide. Thus, in such a conformation, it would be easier to bring the termini together, leaving this rigid "core" of the peptide intact, rather than "unfolding" and reaching a compact state by gathering the ends from the central region of the peptide, as if gathering the ends of a loose rope. The latter mechanism is favored by the presence of the disulfide bond, as it seems to make the peptide more flexible. This unfolding path is purely speculative, as there is no native helical conformation, but as will be seen in Section 3.1.1.2, the cystine moiety does seem to be more flexible, from conformations that have indeed lost their helical content through a short MD run in vacuo at 1000 K. To see if there is a qualitative effect of the charged termini in the end-to-end interactions, a completely random conformation of hIAPP was capped and simulated for at least 50 ns. These data, shown in Figure 3.1b, yield values of R_g that are indistinguishable from the data after the dissolution of the initial α-helical conformation. On the other hand, r_{eted}, in Figure 3.1d, shows comparable values to the uncapped data only in the first portion of the simulation, while the uncapped moiety displays a significant bias towards smaller values of r_{eted} in the second half of the simulation. The uncapped polypeptide does seem to explore the same states as the capped one, albeit less frequently. Further investigation on the charge effect of the uncapped peptide is necessary, since it may introduce bias to at least one of the statistical properties. A deeper analysis on the sampling is carried out in Section 3.1.1.3.

3.1.1.2 Comparison of Independent Starting Conformations

How does the starting conformation affect these statistical properties? Parallel runs on independent random conformations may shed light on this matter. Owing to the strong influence that the secondary structure may have on R_g and possible bias of the α-helical reference conformation, the data were collected from arbitrary initial conformations, labeled with the subscript I, and after the dissolution of the helical elements assigned via DSSP-cont, subscript D.[77,78] Since the initial conformations of RUNS 1 through 4 were selected from families of structures using the GROMOS clustering algorithm[91] on two 1 ns runs at 1000 K in vacuo and subsequently solvated in SPC/E water,[60] as seen in Section 3.1 on page 41, very few data, if any, are removed from the initial run I.* The time dependence of the radius of gyration and end-to-end distance can be seen in Figure 3.2 along with the mean values, also reported in Tables 3.1 and 3.2, in which MD runs for the oxidized hIAPP moiety conformations have been labeled with a prefix Ho, reduced hIAPP with Hr.

The mean R_g values, calculated with `g_analyze`,[80] for all the data sampled at 350 K after complete dissolution of the helical content during $\approx$15 ns runs, shown in the left panel of Figure 3.2b, are within the error bar in four of the five conformations taken from the clustered conformations and are all greater than what is calculated through the random-flight model. There seems to be no apparent correlation between the average helical content and the radius of gyration. The same holds true for the hydrogen bonded turns and bends. Only RUN4 of the oxidized hIAPP moiety seems to have relevant contribution from the secondary structure to minimizing R_g; in fact, the β-structure content, $(5.1 \pm 0.5)\%$, is relatively low, but constant throughout the simulation as can be seen by the small error. At a first glance, it seems that the greater the presence of stable β-structure, the more compact the polypeptide is. Despite having chosen RUN2 and RUN4 of the oxidized moiety from the same structural family, the calculated R_g diverge initially, with RUN2 reaching similar compact conformations at the end of the run, as

---

*Unless stated otherwise, the data gathered after the dissolution of the initial α-helix, D, for RUN1 through RUN4 are labeled I, as the excluded data points, if any, are negligible and do not contribute to the mean or to the error estimate.

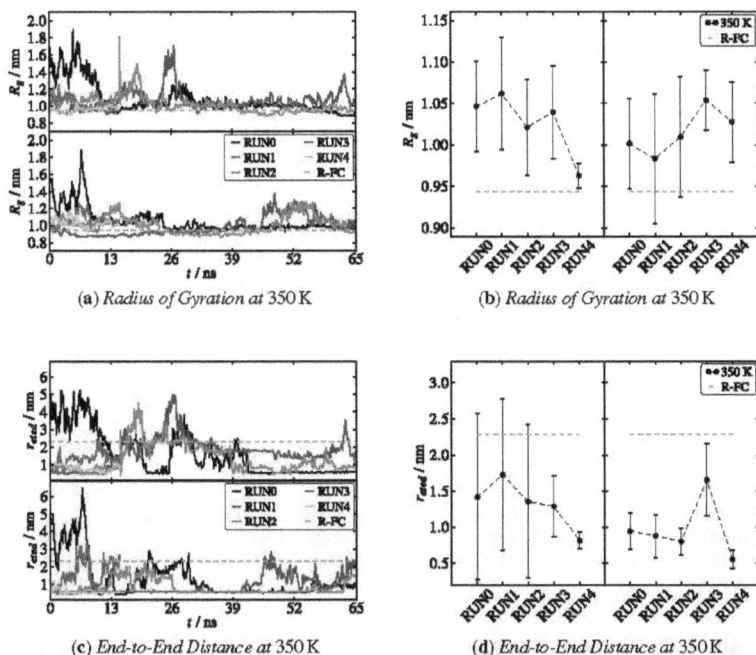

(a) *Radius of Gyration at* 350 K

(b) *Radius of Gyration at* 350 K

(c) *End-to-End Distance at* 350 K

(d) *End-to-End Distance at* 350 K

Figure 3.2: *The time-dependent data relative to oxidized hIAPP are found in the top panels in (a) and (c), while their average values are shown in the left panels in (b) and (d). The reduced hIAPP data are found in the bottom panels in (a) and (c), with the mean values in the right panels in (b) and (d). The mean values shown in (b) and (d) are found in detail in Tables 3.1 and 3.2, in which the prefixes Ho for oxidized hIAPP and Hr for reduced hIAPP, both relative to "dissolved" α-helix, and thus labeled with the subscript D. The dashed cyan line indicates the values calculated for the random-flight chain seen in Table 2.3.*

seen in the top panel of Figure 3.2a. Moreover, looking at the mean values in Table 3.1 closely, HoRUN2 is the only data set that includes HoRUN4. The error of R_g is relatively large, possibly as a result of a temperature effect, but also to insufficient sampling. A deeper analysis is carried out in Section 3.1.1.3.

The five runs for the reduced moiety, as seen in the right panel of Figure 3.2b, all yield comparable average R_g results within the error bar. RUN4 was deliberately chosen for its highly compact state, reached through the simulation in vacuo. Even such structure, if given enough time, is able to be solvated properly and open up to a less compact state. Moreover, this state does not seem to be an artificial or improbable one, as RUN1 also reaches similar compact conformations. These two runs, RUN1 and RUN2, were chosen to be extended by an additional 50 ns, since they had the largest and smallest R_g values, yielding similar average values within the error bar. The longer simulations, RUN0 to

RUN2, present R_g values that are closer to the theoretical one calculated for the random-flight chain model, while RUN3 and RUN4 are both further away from it, perhaps as a result of insufficient sampling, yet still within the error bar of the first three runs.

Table 3.1: *Radius of Gyration at 1 bar and 350 K*

	$\langle R_g \rangle$ / nm		$\langle R_g \rangle$ / nm
HoRUNO$_I$	1.08±0.14	HrRUNO$_I$	1.05±0.10
HoRUNO$_D$	1.05±0.05	HrRUNO$_D$	1.00±0.05
HoRUN1$_I$	1.06±0.07	HrRUN1$_I$	0.98±0.08
HoRUN2$_I$	1.02±0.06	HrRUN2$_I$	1.01±0.07
HoRUN3$_I$	1.04±0.06	HrRUN3$_I$	1.05±0.04
HoRUN4$_I$	0.963±0.018	HrRUN4$_I$	1.03±0.05
HoRUN4$_D$	0.963±0.015		

Table 3.2: *End-to-End Distance between C_α at 1 bar and 350K*

	$\langle r_{eted} \rangle$ / nm	$\langle R_g \rangle \cdot 0.41^{-1}$ / nm		$\langle r_{eted} \rangle$ / nm	$\langle R_g \rangle \cdot 0.41^{-1}$ / nm
HoRUNO$_I$	1.6±1.2	2.6±0.3	HrRUNO$_I$	1.3±0.6	2.6±0.2
HoRUNO$_D$	1.4±1.2	2.56±0.12	HrRUNO$_D$	0.9±0.3	2.43±0.12
HoRUN1$_I$	1.7±1.1	2.58±0.17	HrRUN1$_I$	0.9±0.3	2.4±0.2
HoRUN2$_I$	1.4±1.1	2.49±0.15	HrRUN2$_I$	0.80±0.19	2.46±0.17
HoRUN3$_I$	1.3±0.4	2.54±0.15	HrRUN3$_I$	1.7±0.5	2.56±0.10
HoRUN4$_I$	0.81±0.11	2.35±0.04	HrRUN4$_I$	0.55±0.13	2.51±0.12

Since the R_g data seem to represent the theoretical values of the random-flight chain, they might also describe the expected values seen in Table 3.2 calculated from Eq. (2.10) on page 32. Unfortunately, the discrepancy is pretty large, with the theoretical value being ≈ 2.5 nm and the r_{eted} no greater than 1.7 nm, with the average value oxidized moiety larger than the reduced one. In fact, as seen in Section 3.1.1.1, the end-to-end distance for the cystine moiety values fluctuate much more than the cysteine moiety, suggesting that it is a more flexible biopolymer. Whatever seemed to bias the data towards smaller r_{eted}, is much less evident when comparing the results of the parallel runs performed on the two moieties. In fact, all three of the 65 ns runs for the oxidized moiety, seen in the top panel in Figure 3.2c, show great fluctuations and explore many states, suggesting that the entropic contribution to the Gibbs Free Energy is greater, making it the more stable peptide of the two moieties. What seems to be perplexing though is that a cross-linking disulfide bond normally decreases the conformational entropy* of a disordered peptide; although a disulfide bridge in an ordered conformation stabilizes it, by destabilizing the disordered conformations.[5] The natural disulfide bond, between C2 and C7, may keep the N-terminus ordered, allowing the C-terminus to explore more conformations.

Studies on amylin and its processing intermediates and the effect of the disulfide

---

*The loss of conformational entropy for cross-links connecting two atoms separated by n residues is

$$\Delta S_{conf} = -b - \frac{3}{2} R \ln n,$$

where b may vary from 2.1 to 7.9 cal mol^{-1} °C^{-1}.[5]

bond by Yonemoto et al. show that amyloidosis for amylin free acid[*] is significantly slower than for the capped wild-type amylin,[95] which could be caused by the limitation of configuration sampling as a reult of the formation of salt-bridges. In fact, the average distance between the charges is less than, but not limited to, 0.4 nm, which is the minimum distance to define a salt-bridge in a protein.[96] RUN3, as seen in green in the lower panel of Figure 3.2c, explores many conformations in which such salt-bridge breaks and reforms for a maximum of ≈ 2 ns. Many other runs also present fluctuating r_{eted}, but through the formation of salt-bridges, it seems the flexibilty of the peptide is more limited, especially for the reduced moiety. A deeper analysis on the distribution of the r_{eted} values is discussed in Section 3.1.1.3.

3.1.1.3 Independent Concatenated Data

The question on the amount of data sampled is always a difficult one to answer, especially at lower temperatures, where the sampling is even more limited, let alone calculate its uncertainty. Owing to statistical inefficiency, as already discussed in Section 2.4.4.1 on page 34, it is not possible to obtain the standard error of the mean simply by dividing the standard deviation by the square root of the data points, because MD simulation data are strongly correlated with each other.[45,87] One way to obviate this is to concatenate parallel simulation runs, but it is clear that not all the data are suitable for this. A proper error estimation can help determine which data are good and which are not. Obviously 15 ns, or even 65 ns, are too few, let alone after discarding data; yet it is possible to see, as a first approximation, what the correlation times may be and what is necessary to equilibrate the system in order to collect data in an appropriate production run. Moreover, through the concatenation of the data, it should be possible to see if the parallel runs are equilibrated and have reached a state truly independent from the starting conformation.

Table 3.3: *Standard Deviation vs. Error Estimate*

	$\langle R_g \rangle$ / nm	$\sigma_{\langle R_g \rangle}$ / nm	$\delta_{\langle R_g \rangle}$ / nm	$\langle r_{eted} \rangle$ / nm	$\sigma_{\langle r_{eted} \rangle}$ / nm	$\delta_{\langle r_{eted} \rangle}$ / nm
Ox. hIAPP	1.05	0.14	0.03	1.5	1.0	0.3
Red. hIAPP	1.02	0.12	0.03	1.0_1	0.8	0.17

Table 3.3 contains the mean values of R_g, on the left, and r_{eted}, on the right, of the concatenated runs RUN0, RUN1, RUN2, RUN3, and RUN4, with the relative standard deviation, in the second column of each set, and the error estimate, in the third column of each set.[†] As seen in Tables 3.1 and 3.2, the mean values of the independent runs is comparable in at least three out of five runs for each property of each moiety, but the uncertainty is high. The standard deviation of the concatenated data for both properties for each moiety

---

[*]The ablation of the C-terminal G38 in amylin free acid by the peptidyl amidating monooxygenase (PAM) complex is the last step that yields the amide capped native amylin and glyoxylate.

[†]By combining the average means and standard deviations of independent runs, the same exact results are obtained as concatenating the data and calculating mean and standard deviation on them. It is quicker and less tedious than performing concatenations and all the data crunching, although the error cannot obviously be estimated. All that is needed from each independent set i is the standard deviation, s_i, the size of the set, n_i, and the mean, m_i. The standard deviation of the combined sets is given by $S = \sqrt{\frac{B - \frac{A^2}{N}}{N-1}}$, where $A = \sum n_i \cdot m_i$, $B = \sum \left[s_i^2 \cdot (n_i - 1) + \frac{A_i^2}{n_i} \right]$ and $N = \sum n_i$.[97]

is comparable with the error of RUN0, which is not surprising, since MD runs should sample more conformations starting from an α-helical conformation rather than already compact conformations. If an arbitrary set of data, SET1, has highly correlated conformations that are very different from another arbitrary set SET2, the standard deviation is large because the two sets show fluctuations around two quite different states and can thus give only a few blocks of uncorrelated data, yielding, in extreme cases, a large error. As Eq. (2.16) on page 35 states, the error decreases by increasing the number of blocks of uncorrelated data, hence if the conformations, collected from RUN0 to RUN4, are uncorrelated, but from equilibrated systems, they should contribute to lowering the error. In fact, as seen in Table 3.3, the high standard deviation describes high fluctuation, but the low error confirms a more precise mean value, which is not as trivial as it might seem.

The next step to verify proper sampling of the data is to compare the distribution of the instant values of the time-dependent point, as can be seen in the lower panels of Figure 3.3. The data have been binned according to *Scott's Choice*,[89] as described in Section 2.4.4.3.

The frequencies of values of r_{eted} can be fitted to the radial distribution function, Eq. (2.12) on page 32, or other exponential functions, as has been done by Krukau et al. for the elastine-like peptide GVG(VPGVG)$_3$.[41] The nonlinear curve used for fitting is

$$P(x) = Ax^{\alpha}e^{-Bx^{\beta}}, \tag{3.1}$$

where A, B, α, and β are fitting parameters.* The mean values seen in Table 3.3 are used to calculate the radial distribution of a hypothetical random-flight chain, $W(r_{eted})$, depicted in dashed lines in Figures 3.3c and 3.3d. The outcome of the fitting procedures on the r_{eted} and R_g distributions in Figure 3.3 are detailed in Tables 3.4a and 3.4b, respectively. Besides the previously mentioned fitting parameters of Eq. (3.1), Table 3.4 also contains χ^2, which needs to be minimized to achieve least-squares fitting of the data frequency to the chosen parent distribution; the correlation coefficient, r_{PD}, which describes the correlation between the predicted data, P, and the measured data, D; the root-mean-square of the relative error between these two sets, $\delta_{(P,D)} = \sqrt{\frac{1}{T}\sum_{i=1}^{T}\frac{(P_i - D_i)^2}{D_i^2}}$; and Theil's U_1,[98] which determines how well a model can predict the sampled data, expressed by the following formula:

$$U_1 = \frac{\sqrt{\frac{1}{T}\sum_{i=1}^{T}(P_i - D_i)^2}}{\sqrt{\frac{1}{T}\sum_{i=1}^{T}P_i^2} + \sqrt{\frac{1}{T}\sum_{i=1}^{T}D_i^2}}, \tag{3.2}$$

where U_1 is bound between 0 and 1, with the lower values yielding a better prediction.

As previously mentioned, r_{eted} shows larger fluctuation for the oxidized moiety; in fact, the contour of the frequency (solid line in the lower panel of Figure 3.3c) is fitted by the radial distribution curve relatively well, if not for the peak on the left ($r_{eted} \approx 0.5\,\mathrm{nm}^{\dagger}$) given by the apparent bias towards minimum end-to-end distance. This bias influences

---

*All of the data presentation has been performed by ad hoc python scripts written with MatPlotLib, SciPy, and NumPy packages, although the curve_fit function of scipy.optimize did not seem to fit the R_g in any way, so the fitting was performed by calling the nonlinear least-squares curve fitting of gracebat (available at http://plasma-gate.weizmann.ac.il/Grace/).

†This peak of short r_{eted}, calculated between C_{α}, corresponds to a salt-bridge between the uncapped C-terminus and either charge of K1, as seen in Section 3.1.1.2.

RANDOM CONFORMATIONS FROM VACUUM

Table 3.4: *Nonlinear Curve Least-Squares Fit*

	(a) Nonlinear Curve Fit $P(R_g)$					(b) Nonlinear Curve Fit $P(r_{eted})$			
	4 param.		2 param.			4 param.		2 param.	
	Ox.	Red.	Ox.	Red.		Ox.	Red.	Ox.	Red.
A	63.8	20.3	32.3	40.1_7	A	1.60	2.36	2.02	19.7
B	2.5	1.6	2.53	2.72	B	1.18	1.55	1.16	4.21
α	25.3	11.8	2	2	α	0.40	0.31	2	2
β	9.9	8.6	2	2	β	0.98	2.18	2	2
χ^2	13.37	19.65	118.9	70.49	χ^2	2.88	22.1_1	4.23	20.3
r_{PD}	0.98	0.97	0.78	0.86	r_{PD}	0.73	0.61_5	0.61_1	0.66
RMS $\delta_{(P,D)}$	-	3.29	-	7.85	RMS $\delta_{(P,D)}$	3.17	1.47	0.956	1.15
U_1	0.10	0.11_8	0.31_8	0.23	U_1	0.31_9	0.47	0.40	0.43

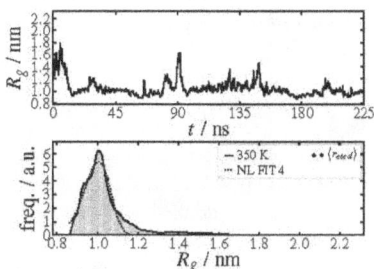

(a) *Radius of Gyration at* 350 K

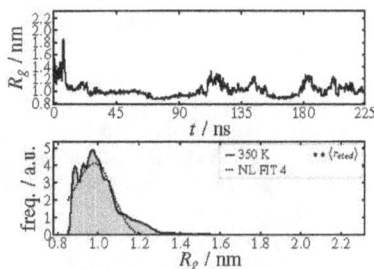

(b) *Radius of Gyration at* 350 K

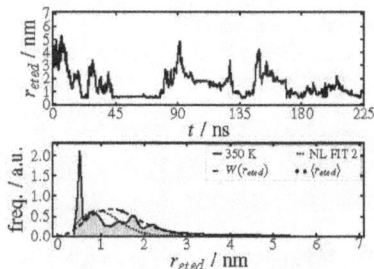

(c) *End-to-End Distance at* 350 K

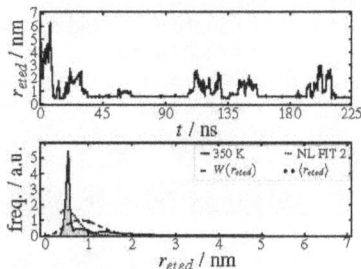

(d) *End-to-End Distance at* 350 K

Figure 3.3: *The time-dependent data in the top panels are relative to the concatenation of RUN0 through RUN4 with a frequency distribution in the lower panels, where an attempt of non-linear curve fitting with two or four terms was performed (indicated with a dotted line; labeled NL FIT 2 or 4, respectively). Also in the distribution, points indicate the position of the mean value seen in Table 3.3 relative to the fit. (a) and (b) show R_g for the oxidized and reduced hIAPP moiety, respectively. In (c) and (d) the radial distribution, Eq. (2.12) on page 32, is also shown with dashed lines.*

the fitting of the non-linear curve (Figure 3.3c, dotted line lower panel) shifting it towards smaller values of r_{eted}, albeit maintaining a shape comparable to the theoretical distribution $W(r_{eted})$ (Figure 3.3c, dashed line lower panel). Obviously, the fit for the oxidized moiety is better than that of the reduced one, but it still must be taken with a grain of salt and should be confirmed by visual inspection of the data.

Table 3.5: *End-to-End Distance from Fitting*

	$\langle r_{eted}\rangle$ / nm	$\langle r_{eted}\rangle_A$ / nm	$\langle r_{eted}\rangle_B$ / nm
Oxidized hIAPP	1.5±0.3	0.9±0.8	1.1±1.1
Reduced hIAPP	1.01±0.17	0.4±0.5	0.6±0.7

The best fit for r_{eted} is given by setting α and β to two, while allowing A and B to be optimized. The χ^2 value for oxidized IAPP in Table 3.4b for the two parameter fit is greater than the one for four (4.23 vs. 2.88) and the slightly lower U_1 value would suggest that the four parameter fit is better, but the RMS of the relative error is much lower for the two parameter fit (0.96 vs. 3.17), besides being the most relevant as it corresponds to the radial distribution seen in Eq. (2.12) on page 32. In fact, both fitting parameters A and B can be used to calculate $\langle r_{eted}^2 \rangle$ by comparing Eq. (2.12) and Eq. (3.1), B is equal to β^2 and A is proportional to β^3. The relative error in this fitting is at least 0.96, so the result is not significant, but, in the best case, it could yield a supplementary check on the goodness of fit; in fact, the oxidized moiety data sampling leads to a fitted $\langle r_{eted}\rangle$ value closer to the measured one, as found in the first row of Table 3.5. In fact, the reduced moiety presents a more prominent bias towards shorter end-to-end distances, as seen in both the lower panel of Figure 3.3d and the output of the fitting in Table 3.4b. As a result, both the fitting curve and the radial distribution $W(r_{eted})$ are markedly shifted towards this maximum in the distribution, as can be seen by the χ^2 values of 20 or more.

There seems to be no specific probability function capable of describing the distribution of R_g, therefore all four parameters should be fitted. Although it is possible to set α and β to two as for the end-to-end distance, the results are terrible; in fact, the χ^2 are at least 70, as seen in Table 3.4a. Thus, all four fitting parameters were used to fit the R_g distribution in Figure 3.3a and Figure 3.3b, but once more the fitting is not good; in fact, the shoulder on the right side of both distributions is poorly fitted by this nonlinear curve, therefore, such an analysis on R_g will not be investigated further.

3.2   Extended Trajectories

Collecting significant data for the test runs, illustrated in the previous sections, was relatively successful, considering the brevity of those MD simulation runs. Said success can be partially attributed to the relatively high temperatures used, especially for physiological temperatures. The starting conformation of each run at multiple temperatures, from 250 K to 450 K every 20 K, was the last conformation of the pre-equilibration run of 15 ns RUN0, which started from an α-helical conformation (ALPH), as described in Section 2.2.2 on page 16. The box-size at 350 K for the cysteine moiety was 6.989 70 nm, containing 10 842 water molecules, and 6.989 18 nm for the cystine moiety, with 10 843 water molecules. Such conformation was heated, or cooled, by running a 2 ns equilibration run starting at 350 K to the desired temperature, with the same conditions previously applied. Upon equilibration of the density of the system, these conformations were all then simulated for 250 ns discarding the first 50 ns as pre-equilibration. In order to allow the polypeptide to sample the conformational space at lower temperatures, it is necessary to perform longer runs and possibly concatenate equilibrated conformations. Owing to this difficulty, five initial conformations were chosen from MD simulations performed at 350 K and 450 K, with at least 10 ns or 50 ns of data discarded as pre-equilibration, as a result of long correlation times. Besides the ALPH conformation, four other initial conformations were chosen, namely, RTSS, a conformation characterized by a positive (right-handed) value of $\tau(S-S)$ obtained from the last conformation of the ALPH trajectory at 350 K after 265 ns simulation, and three conformations CL01, CL02, and CL03, obtained from clustering the ALPH trajectory at 450 K as follows: a trajectory of 50 ns that presented the largest fluctuations of r_{eted} at 450 K was chosen from the complete 200 ns trajectories. Hence, the first set of 50 ns of the production run was chosen for the reduced moiety, while the third set was preferred for the oxidized moiety, as it presented a broader distribution of r_{eted} than the first one. Even though a simulation in vacuo at 1000 K is faster than a simulation in water at 450 K as seen in Section 2.2.1, sampling of "random" or arbitrary conformations was carried out for a solvated system in order to reduce the electrostatic interactions of the charged termini. The GROMOS clustering algorithm[91] with an all-atom 0.9 nm RMSD cutoff was performed on either trajectory taking every other structure and subsequently choosing the middle structure of the most represented conformation for CL01, the least represented for CL03, and an arbitrary one in the middle for CL02. Surprisingly, the reduced moiety had more clusters (17) than the oxidized moiety (10), even though it seems that at lower temperature the sampling is favorable for the oxidized moiety. Of the 12 501 structures, 66 % comprise the main cluster for oxidized hIAPP and 53 % for the reduced moiety, while 1.5 % (ox. hIAPP) and 1.1 % (red. hIAPP) comprise the arbitrary middle cluster, and 0.03 % and 0.02 % comprise the last cluster, oxidized and reduced hIAPP respectively. Conformations of proteins can be thought of as a range of closely related microstates, which can be visited rapidly by the protein from one to another, while at lower temperature it can be "trapped" in one microstate or another. The Free Energy for these states was not calculated, but it is possible to estimate that the most frequented states are the ones with the least energy, while the ones with least occurrence have the most energy.[5] CL02 and CL03 were purposely chosen to investigate the temperature effect on conformations that were highly improbable at lower temperatures. In other words, at what temperature would the water shell have to be for it to be "soft" enough to allow the peptide to return within equilibrated values for the various properties, e.g., R_g and r_{eted}. Runs of 60 ns were performed on these initial conformations, discarding the first 10 ns, at temperatures ranging from 250 K to 350 K, every 20 K, as can be seen in Figures 3.4 and 3.5.

In Figures 3.4a and 3.4c, the mean values of R_g and r_{eted} for the single 200 ns run ALPH are compared to the values of the concatenated runs ALL2, ALL3, ALL5, where the additional runs RTSS, CL01, CL02, and CL03, are subsequently concatenated to ALPH in the following manner: 150 ns of RTSS are concatenated to ALPH in all three cases, ALL3, includes 50 ns of CL01, and ALL5, also includes 50 ns of CL02 and CL03. The mean values of R_g and r_{eted} of the individual runs, RTSS, CL01, CL02, and CL03 can be seen individually in Figures 3.4b and 3.4d. At a first glance, the data relative to oxidized hIAPP (left panel in all subfigures of Figure 3.4) are more homogeneously spread and seem readily equilibrated, compared to the reduced moiety (right panels,

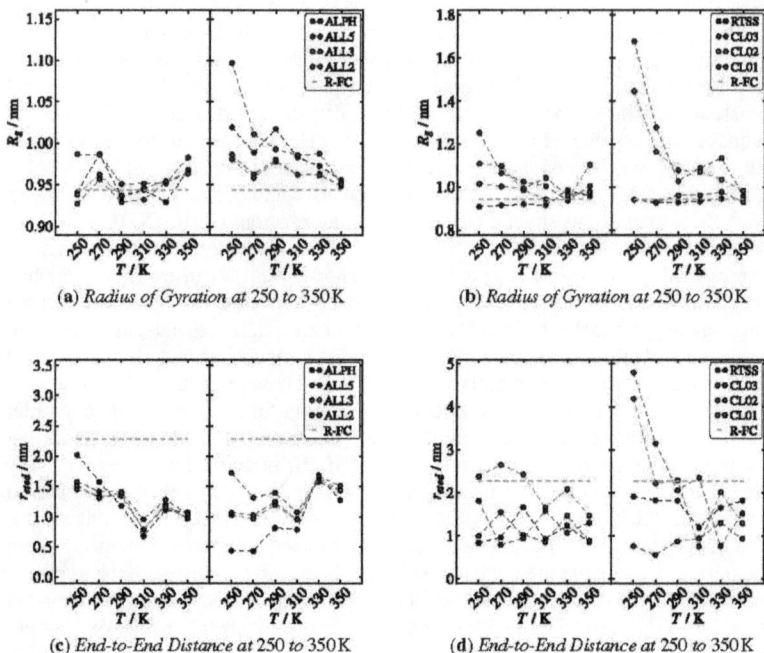

(a) *Radius of Gyration at* 250 *to* 350 K

(b) *Radius of Gyration at* 250 *to* 350 K

(c) *End-to-End Distance at* 250 *to* 350 K

(d) *End-to-End Distance at* 250 *to* 350 K

Figure 3.4: *The mean values of the statistical properties (a) R_g and (c) r_{eted} for the single run ALPH and the concatenated runs ALL2, ALL3, and ALL5 are plotted versus temperature, with (b) R_g and (d) r_{eted} showing the individual runs, RTSS, CL01, CL02, and CL03, which have been concatenated to ALPH. The theoretical values of R_g and r_{eted} for a random-flight chain are depicted by a dashed cyan line. In all the figures, oxidized hIAPP is in the left panel, while reduced hIAPP is on the right.*

Figure 3.4), especially at lower temperature; in fact, both Figures 3.4b and 3.4d have different ordinate scales than Figures 3.4a and 3.4c. If the R_g calculated for the random-flight chain is taken as a reference, both ALPH and ALL5 of the oxidized moiety (left panel Figure 3.4a) seem to deviate from it at lower temperature and at 350 K. This could mean that the ALPH run at 270 K was not properly sampled and that RTSS and CL01, taken from trajectories at higher temperature, contribute to average out R_g. This is particularly true for the reduced moiety (right panel Figure 3.4a), where ALL2 and ALL3 are both closer to the expected theoretical value (cyan dashed line). Evidently, the contribution of the higher temperature used to obtain CL01 might have helped explore more compact states, which for the cysteine moiety are insufficiently sampled at lower temperatures. The long simulation time required to find the initial RTSS conformation might have compensated for the low temperature, helping compact the overall structure of the sample conformations. So it seems that for the runs to yield comparable values of R_g, the temperature should at least be 290 K; in fact, even though the initial values are quite far from the theoretical values, above 270 K the values become more and more comparable to the theoretical value. Similar observations can be made for r_{eted}, as CL03 for the reduced moiety is clearly far from equilibrium at 250 K and 270 K, as is CL02 at 250 K, as can be seen in the right panel of Figure 3.4d, while the oxidized moiety presents average values of r_{eted} that are all comparable, with the exception of CL02, found to be slightly above the expected value for a random-flight chain, as seen in the left panel of Figure 3.4d. The average values of r_{eted} for the concatenated values of the reduced moiety also confirm that ALPH and ALL5 diverge from ALL2 and ALL3 at lower temperature and reach comparable values at 310 K (right panel Figure 3.4c), while the oxidized moiety presents similar values for all concatenated values (left panel Figure 3.4c).

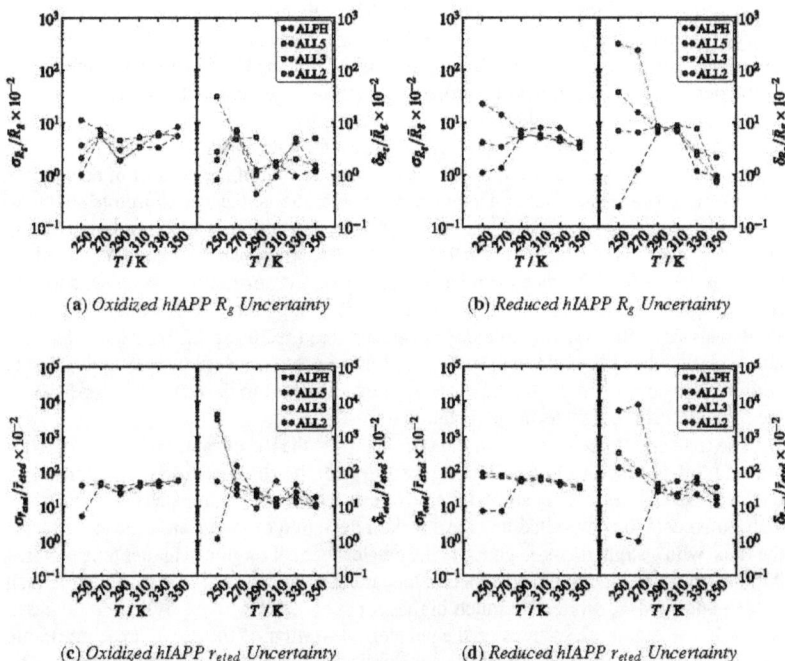

(a) *Oxidized hIAPP R_g Uncertainty*   (b) *Reduced hIAPP R_g Uncertainty*

(c) *Oxidized hIAPP r_{eted} Uncertainty*   (d) *Reduced hIAPP r_{eted} Uncertainty*

Figure 3.5: *Temperature-dependence of the relative standard deviation of the mean (left panels) and relative error (right panels) for the concatenated data ALL2, ALL3, ALL5, and the individual run ALPH. Owing to the amplitude of the relative estimated error at lower temperature, the data are represented in semi-logarithmic scale on the ordinate axis. (a) and (b) are relative to R_g and (c) and (d) are relative to r_{eted}, with the oxidized hIAPP on the left and the reduced one on the right.*

In order to decide which concatenated data should be chosen between ALL2 and ALL3, it is necessary to look at the estimated errors of both properties for both moieties and not only their average values. In Figure 3.5, the left panels show the standard deviations of the mean of the concatenated data, which are, as previously mentioned in Section 3.1.1.3, identical to the combined standard deviation of the independent runs,[97] while the right panel shows the relative error obtained through g_analyze.[80] When the data are correlated within the sets and the sets are quite different from each other, the standard deviation is large, and the error is even larger. As can be seen in the left panels of Figure 3.5, the standard deviation for ALL5 is larger or equal to the other runs, which is not surprising, since the initial conformations of CL02 and CL03 constituted only 1.5% or less of the 50 ns trajectories sampled at 450 K. Owing to the huge errors, especially at lower temperature, it was necessary to plot the ordinate in logarithmic scale. Please note that the relative error for r_{eted} is two orders of magnitude larger than that of R_g. There is no clear trend of the temperature dependence of the errors compared to the standard deviation, but it does seem that in the runs above 330 K, and in some cases above 290 K, the error is significantly lower than the standard deviation, so the data of the concatenated sets are sufficiently uncorrelated and thus participate in the data sampling of similar states, contributing to a "truer" mean. Obviously, given the length of the peptide, its structureless nature, and the strong interactions between its residues, the sampling in MD simulations becomes even harder, so a "true" mean, especially at lower temperatures, is Utopian. In other words, even running 1000 ns simulations would not lower the error by a factor of $\approx$700, i.e., the square root of the number of data points, let alone at lower temperatures. Upon these considerations, the sets chosen to study the water shell properties are ALL3, CL02, and CL03. Even though ALL3 presents a large error in R_g for the reduced moiety at 250 K and 270 K (right panel Figure 3.5b, in red), it was chosen over ALL2, since the relative error of r_{eted} is smaller for the reduced hIAPP (red vs. green in the right panel of Figure 3.5d). The data used for the initial water shell analysis in Ref 43, available in Appendix A, are the 200 ns ALPH trajectories, and although they have been proven, in this section, to contain insufficient sampling of the conformational properties, the trajectories were sufficient to describe the properties of the hydration shell, such as the percolation transition.

The temperature dependence of the uncertainty of the mean values of R_g and r_{eted} for the ALL3 data is shown in detail in Figure 3.6, with the standard deviation depicted by the error bars in the left panels and the error estimate in the right panels. For normally distributed data, it is expected that the standard deviation of the mean is the same for all the runs, with a slight increase given by the greater thermal motion at higher temperatures. As one can see in Figure 3.5, the relative standard deviation of the mean for R_g is between 2.96% and 5.94%, while it is much higher for r_{eted}, 24.4% to 56.5%. In both cases, for each set of data, the values of the standard deviation of the mean are comparable within each set. In the right panel of each subfigure, the estimated error is shown. The oxidized moiety shows a significant reduction of the error from the standard deviation of the mean starting from 290 K for R_g (right panel Figure 3.6a) and from 310 K for r_{eted} (right panel Figure 3.6c). Obviously, the error for r_{eted} at 250 K is enormous, as seen in Figure 3.5c. Unfortunately, the data for reduced hIAPP is not quite as good. The errors for R_g at low temperature are even larger than those for the oxidized moiety, as shown in Figure 3.5b, and are smaller than the standard deviation only at 330 K and 350 K (right panel Figure 3.6b). But then again, the ALL3 concatenation was chosen because the error was smaller for r_{eted}; in fact, the error is smaller than the standard deviation starting from 290 K (right panel Figure 3.6d). Moreover, the low sampling at lower temperature

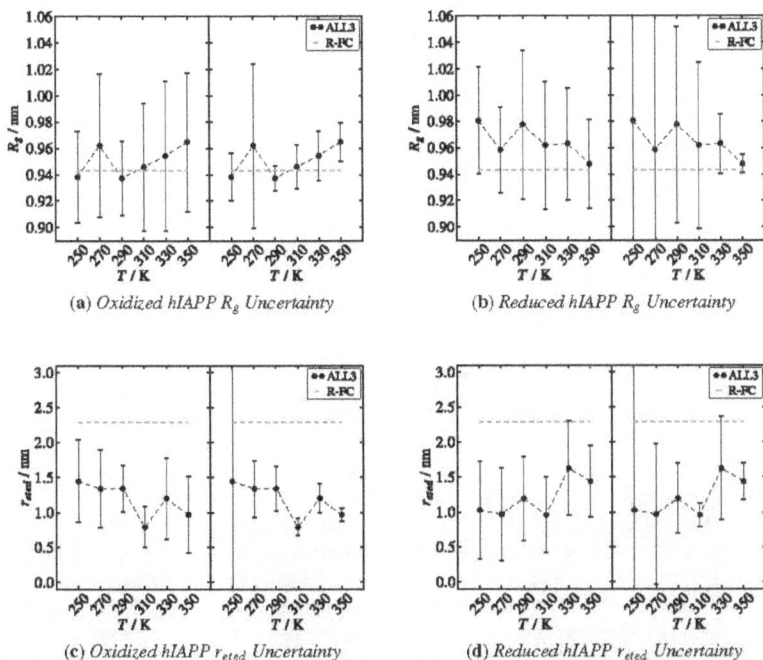

(a) *Oxidized hIAPP R_g Uncertainty*

(b) *Reduced hIAPP R_g Uncertainty*

(c) *Oxidized hIAPP r_{eted} Uncertainty*

(d) *Reduced hIAPP r_{eted} Uncertainty*

Figure 3.6: *Temperature dependence of the mean values of R_g ((a) for the oxidized moiety and (b) for the reduced moiety), and r_{eted} ((c) for the oxidized moiety and (d) for the reduced moiety) of the concatenated run ALL3, where the error bars in the left panels represent the standard deviation of the mean, while those in the right panels represent the estimated error. The large error bars were truncated, in order to see the trends of the standard deviation error of the mean vs. the error estimate. For a complete representation of the error estimate, please see Figure 3.5, which is a semi-logarithmic plot.*

yields, once again, a very large error at 250 K.

Comparing these data to the theoretical values for a random-flight chain (Figure 3.6 dashed cyan line), the mean value of r_{eted} is smaller for both moieties (Figures 3.6c and 3.6d), while R_g is larger for the reduced moiety (Figure 3.6b) and fluctuating about it for the oxidized moiety (Figure 3.6a).

Although there are some uncertainties with the data relative to mean values of R_g for the reduced hIAPP moiety, an interesting trend can be seen in Figure 3.7; in fact, the large error at 290 K and 310 K is caused by the presence of two distinct states, as can be seen in the lower panel in Figure 3.7b. At these two temperatures (distributions in black for 290 K and blue for 310 K), the peaks around 1.05 nm are highly populated and do not seem to merge with the other data, resulting in highly populated peaks that contribute to a large error since the two sets are nicely distributed individually, but do not mix.

Upon heating, these peaks that correspond to a larger R_g, become nothing more than a tail of an asymmetrical distribution (distributions in red for 330 K and green for 350 K), hence the error becomes smaller than the standard deviation of the mean since the data are uncorrelated. Whether this can be caused by the percolation transition at $\approx$320 K or simply greater flexibility of the peptide and higher kinetic energy because of a higher temperature is still to be determined, but it is interesting, nevertheless, to notice such a coincidence. The error for the oxidized moiety is small for all the temperatures, because the data are more compatible, and there is only a slight shoulder in the distribution, as can be seen in black for 290 K and blue for 310 K in the lower panel in Figure 3.7a, which is not as populated as what was seen for the reduced moiety. Unfortunately, there is no clear-cut trend for the r_{eted} distribution, also confirmed by the fact that many errors between 290 K and 350 K illustrated in Figures 3.6c and 3.6d are of the same order of magnitude of the standard deviation of the mean. Therefore, the data are not bad, because comparable to the standard deviation of the mean, but not sampled well enough to reduce the error. A trend that can be seen for the reduced moiety is the lowering of the peak around 1.8 nm upon heating, while the distribution for the oxidized moiety is the shift that occurs for the entire distribution from 290 K to 310 K to lower values of r_{eted}. Even though from 10 ns to 50 ns data were discarded as pre-equilibration, please note that all these trajectories, ALPH, RTSS, and CL01, were obtained from the same starting conformation, so one can see the temperature effect on properties such as R_g and r_{eted}. Obviously, discarding, for example 10 ns, from five parallel runs at different temperatures, will not give the same initial conformation for the production run, but it could sometimes be correlated, as can be seen for r_{eted} of the reduced moiety (Figure 3.7d).

From what has been thus far demonstrated, the data at 350 K seems to present sufficient sampling to suppose that only one run at higher temperatures, i.e., from 370 K to 450 K, should suffice to describe trends of temperature dependence for conformational properties; in fact, such data are useful for providing a more complete picture on the hydrational properties of the peptide. Higher temperatures are not considered, especially because they are not physiologically feasible considering that peptide bonds are hydrolyzed in $\approx$1 min at 250 °C.[5]

3.3   Conclusions

In this chapter, a detailed description has been presented on the preparation of hIAPP for the study of conformational properties, secondary structure, and hydrational and volumetric properties of the first layer of the hydration shell. In order to simplify the system as much as possible, the positive charge of the polypeptide determined by the N-terminus, lysine-1, and arginine-11 has been scaled to neutrality by distributing an equal and negative charge amongst the residues. This leads to slightly more negative partial charges and less positive charges by less than 1.5 % of the original value. As far as can be seen by the graphical analysis of Bland-Altman plots, a possible systematic error of 1.5 % on the electrostatic potential given by the scaled partial charges has been introduced by this procedure, although it should not influence the determination of the secondary structure as the torsional terms in the OPLS-AA are determined through ab initio calculations. The Ramachandran angles that have been chosen to encompass regions assigned to helical and extended conformations do not seem to be influenced by the scaling of the charges, if not

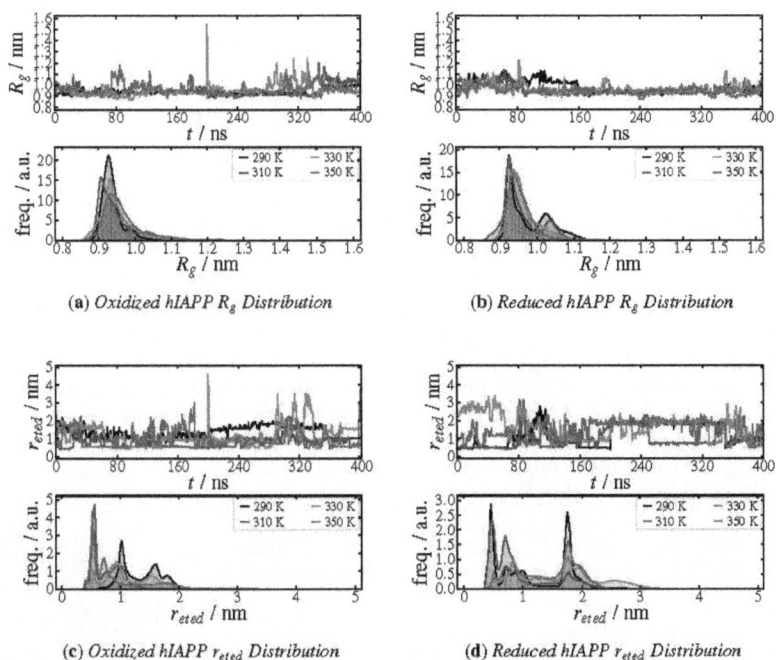

(a) *Oxidized hIAPP R_g Distribution*

(b) *Reduced hIAPP R_g Distribution*

(c) *Oxidized hIAPP r_{eted} Distribution*

(d) *Reduced hIAPP r_{eted} Distribution*

Figure 3.7: *The time-dependent data relative to temperatures* 290 K *to* 350 K *resulting from concatenation of ALPH, RTSS, and CL01 are seen in the top panels, while their frequency distributions are depicted in the lower panels.*

by a slightly sharper peak in the $60° \times 60°$ region that defines the helical conformations. Moreover, the helical contribution of the polypeptide to the overall secondary structure seems to correspond well to secondary structure assignment software, albeit with a slight overestimation since the definition of the Ramachandran angles does not consider other properties to determine helices, such as H-bonds, nor does it assign secondary structure to a minimum number of consecutive helical elements. Hence, a single residue may be assigned to a helix, even if it actually is not part of three or five consecutive residues that define a helix, but it can possibly foresee the formation of helices one element at a time, as opposed to seeing the abrupt formation and dissolution of helical elements of three, four, or five residues. As this analysis was performed on short runs to investigate on the effect of the charge scaling and for the selection of the proper Ramachandran angles to determine the regions of the three secondary structures considered, an actual study on the secondary structure of IAPP will be performed in Chapter 5.

The full-length hIAPP was run in vacuo at 1000 K for 1 ns to sample random conformations in order to choose an adequate and unbiased initial conformation for MD simulations in liquid water, although slightly longer runs in water at 450 K give similar results. These trajectories seem to be biased by a charge-charge interaction between the negatively charged uncapped dissociated C-terminus and the positively charged protonated N-terminus, partially limiting the exploration of more extended conformations, therefore investigation on secondary structure should be performed on a peptide bearing a neutrally charged amide-capped C-terminus, as seen in Chapter 5. Of the statistical properties presented in this chapter, the end-to-end distance, r_{eted}, seems to be influenced the most by the presence of the charged termini, while the radius of gyration, R_g, seems to be independent from the charges on the termini. Two forms of hIAPP, characterized by different oxidation states, i.e., with and without disulfide bond between C2 and C7, seem to manifest different flexibility, with the oxidized form seemingly more capable of exploring the conformational space than the reduced counterpart.

The difficulty of sampling of the 37-residue polypeptide at lower temperature was partially circumvented by performing runs from different starting conformations; in fact, at 330 K the collected data can be concatenated and averaged out, which coincidentally takes place after the calculated percolation transition that occurs at approximately 320 K. The MD simulations at lower temperatures, especially below 310 K, were performed mostly in order to collect data for the studies on the hydration shell properties. In particular, a few conformations at 250 K were chosen to evaluate such properties, where the peptide is less flexible without being frozen since the SPC/E water model does not show a density maximum until 234 K.

Chapter 4

Water Percolation

4.1 Hydration Water Properties

Through MD simulations it is possible to study in detail the temperature effect on the connectivity of hydrogen bonds in the hydration shells of islet amyloid polypeptides and other biomolecules. Many variants of IAPP have been studied to see the effect of chemical modifications on the hydrogen-bonded network of hydration water that homogeneously envelopes a peptide at low temperature and breaks into an ensemble of small clusters upon heating. Of the many properties analyzed, the ones that show strong correlation, or rather anti-correlation, are the radius of gyration, R_g, and the solvent accessible surface area, SASA; in fact, these two properties of IAPP start to increase when the hydration water network breaks upon heating. The fluctuations of the number of intrapeptide hydrogen bonds show negative, or anti-, correlation with the fraction of molecules in the largest cluster of hydration water. The more intrapeptide H-bonds formed, the more thermally stable the network of hydration water is, resulting in a more hydrophobic peptide surface. The thermal stability of the H-bonded water network in the hydration shells of the IAPP variants and several other biomolecules is found to be rather similar: the network breaks between 300 K and 330 K, i.e., in the temperature interval where the biological activity of living organisms is maximal. This particular temperature range was chosen to analyze the IAPP variants and, in particular, the oxydized moiety of hIAPP, as Kayed et al. have reported that its experimentally measured lag time in aggregation drops drastically at about 320 K.[10] The conformational changes that have been observed in the MD simulations of IAPP variants are described in Chapter 5.

The properties of water near surfaces differ strongly from the properties of bulk liquid water.[99] At biologically relevant thermodynamic conditions, the surface effects do not spread essentially beyond the first surface water layer. Thus, approximately a monolayer of liquid water adjusted to the surface, hydration water hereafter, can be considered as a subsystem with its own structural, dynamic, and thermodynamic properties. The properties of hydration water can hardly be measured experimentally, but they have been intensively studied near various surfaces by simulations.

In general, the thermodynamic properties of water near surfaces seem to follow the general laws of the surface critical behavior observed for simple fluids and for lattices.[99-101] The density of hydration water decreases almost linearly with temperature upon heating and, at ambient conditions, its thermal expansion coefficient* exceeds that of bulk water near both hydrophobic and moderately hydrophilic surfaces.[102] Owing to the high thermal expansivity,[†] the constant pressure heat capacity of hydration water notably exceeds the bulk value.[42] The presence of the highly directional water-water H-bonds causes additional surface effects specific for H-bonded fluids only. Water properties near a surface are affected by rearrangement of water-water H-bonds as well as by the formation of water-surface H-bonds.

The degree of the *connectivity* of the H-bonds between the water molecules in the hydration shell is an important characteristic of hydration water, which can affect its thermodynamic and dynamic properties.[105,106] Two quite different states of the connectivity of the H-bonds in the hydration shell can be distinguished: a) hydration water forms an infinite percolating, or spanning, H-bonded network; b) hydration water consists of small finite H-bonded clusters. At the surface of a finite object, e.g., a biomolecule, the percolating network is always finite, but its spanning characteristic appears in the homogeneous coverage of a surface.[107] The simulation studies have shown that the hydration water near various bio-surfaces forms a quasi-two-dimensional spanning H-bonded network at low temperatures. Upon heating, the number of water-water H-bonds within the hydration shell decreases and the spanning network of hydration water breaks into an ensemble of small H-bonded clusters.[40,90] This process may be well described as a quasi-two-dimensional percolation transition and it occurs in a biologically relevant temperature range.[40,99] Taking into account the crucial role of hydration water in biology, it is possible to assume that the drastic change in the connectivity of H-bonds influences properties of both hydration water and biomolecules.

Up to now, the thermal break of the H-bonded network in the hydration shell has been studied for several polypeptides in water: elastin-like peptide,[40,41] $A\beta_{42}$-peptide,[22] NFGAIL-peptide and GNNQQNY-peptide.[42] The molecular weight of these peptides varies greatly, from 634 Da (NFGAIL-peptide[‡]) to 4511 Da ($A\beta_{42}$-peptide), with the oxidized moiety of hIAPP being 3906 Da. The temperature interval, where the spanning network of hydration water breaks upon heating, is rather similar in all cases, although the width of the temperature interval where the network breaks is different for the different biomolecules. The water network has been found stronger in the hydration shells of thermophilic proteins,[109] although in that study, Sterpone et al. did not estimate the temperature of the thermal break of the spanning water network. The reason for rather similar stability of spanning water networks in the shells of many biomolecules is not clear. At a first glance, the temperature stability of the spanning network of hydration water may depend on the size of a biomolecule and on the degree of the hydrophilicity/hydrophobicity of the biomolecular surface. Both factors can affect water-water H-bonding within the hydration shell and, accordingly, the thermal stability of the spanning network of hydration water. However, from the data available, it remains unclear how the thermal break of hydration water depends on the properties of a biomolecule. More simulations of biomolecules

---

*$\alpha_h = -\frac{1}{\rho_h}\frac{\delta\rho_h}{\delta T} = \frac{\delta(-ln\rho_h)}{\delta T}$;[102] $\alpha_V = \left(\frac{1}{V}\frac{\delta V}{\delta T}\right)_p$ [103] pg. 659, Berry et al.

†Heat capacity and thermal expansion, as well as elasticity, have been empirically proven to be qualitatively correlated; in fact, elastically stiff materials present low thermal expansion, while materials with high thermal expansion are capable of absorbing more energy per unit temperature increase leading to thermal expansion.[104]

‡This amino acid sequence was studied in virtue of its putative amyloidogenic properties; in fact, it was found to be the shortest fragment of IAPP capable of aggregating.[108]

with various chemical structures and conformations are required to answer this question.

In this chapter, a study on the thermal stability of the hydration water network near several modifications of the islet amyloid polypeptide by simulations is presented. Brovchenko et al. have studied the thermal stability of the hydration water network at the surface of another similar amyloidogenic peptide of approximately the same size as hIAPP, i.e., $A\beta_{42}$-peptide, although the simulations were much shorter (a maximum of 20 ns)[22] and thus presented a rather strong scattering of the data points. The data sampled for IAPP is essentially improved through much longer trajectories (up to 400 ns), allowing greater accuracy of the data. Not only is this data compared to $A\beta_{42}$-peptide, but also to even longer trajectories of IAPP variants that present different chemical structure. The resulting effect of the chemical structure and conformation of the polypeptide on the spanning character of the hydration water network is analyzed and compared to those of the aforementioned biomolecules.

4.2 Hydration Water Analysis

4.2.1 Temperature-Induced Percolation Transition of Hydration Water

The probability distributions of the fraction of the largest cluster relative to the number of water molecules in the hydration shell of hIAPP at various temperatures, $P(S_{max}/N_w)$, are shown in the upper panel of Figure 4.1, with the distributions for other IAPP variants being very similar. The evolution of this distribution with temperature shows the process of the break of the H-bonded network of hydration water. At the lowest temperature studied, the largest cluster of hydration water includes most of the water molecules, as can be seen at 250 K, where the percolation probability, defined by the average value of S_{max}/N_w, is about 0.92. A feature of the percolation probability is that it is roughly symmetrical around 0.5, and for these data, the broadest distribution occurs at 310 K and 330 K, as can be seen by dashed lines in the upper panel of Figure 4.1. This can also be seen in the upper panel of Figure 4.3, where the width $\Delta(S_{max}/N_w)$ of the probability distribution $P(S_{max}/N_w)$ changes non-monotonically with temperature showing a maximum at about 310–330 K. In fact, at low and high temperatures, the distribution $P(S_{max}/N_w)$ is rather narrow, while it is wide at intermediate temperatures of about 310 K to 330 K. The temperature dependencies of the width $\Delta(S_{max}/N_w)$ of the probability distribution $P(S_{max}/N_w)$ for the hydration shells of hIAPP and hIAPP^{S*} are shown in the upper panel of Figure 4.3, in which a maximum at $T = (324 \pm 3)$ K can be seen for both moieties. This means that the strongest variations of the size of the largest water cluster occur at this temperature.

Given the rough symmetry around $S_{max}/N_w = 0.5$ of the probability distributions shown in the top panel of Figure 4.1, the probability that the largest cluster contains more than half of water molecules in the hydration shell is defined as the spanning probability, SP. The spanning probability is calculated as an integral of the probability distributions shown in the upper panel of Figure 4.1 for $S_{max}/N_w \geq 0.5$ and can be seen in the middle panel of Figure 4.2 as a function of temperature. The spanning probability can be fitted to a sigmoidal function with the inflection point at $T = (322.5 \pm 0.5)$ K, corresponding

---

*In this chapter, the oxidized moiety is denoted by the superscript "S" for consistency with the figures. The absence of the superscript refers to the reduced moiety.

to $SP = 0.5$. The temperatures of the inflection point for hIAPP and hIAPP^S differ by less than 0.1 K. Similar to previous studies, the temperature corresponding to $SP = 0.5$ is assigned to the midpoint of the percolation transition of hydration water in the hydration shell of a biomolecule. This midpoint is rather close to the temperatures where the mean cluster size S_{mean} and the width $\Delta(S_{max}/N_w)$ of the distribution $P(S_{max}/N_w)$ are maximal, as can be seen in Figure 4.3.

The temperature dependence of $(S_{max}/N_w)^{av}$ is shown in the upper panel of Figure 4.2. Upon heating, the largest cluster includes less and less water molecules and $(S_{max}/N_w)^{av}$ does not exceed ≈ 0.3 at temperatures above 350 K. Although the percolation probability behaves differently below and above the percolation threshold, [110] the fragment of the temperature dependence $(S_{max}/N_w)^{av}(T)$ can be approximated by a sigmoidal function with the inflection point at about 314 K for both hIAPP and hIAPP^S peptides, depicted by the lines in the upper panel of Figure 4.2.

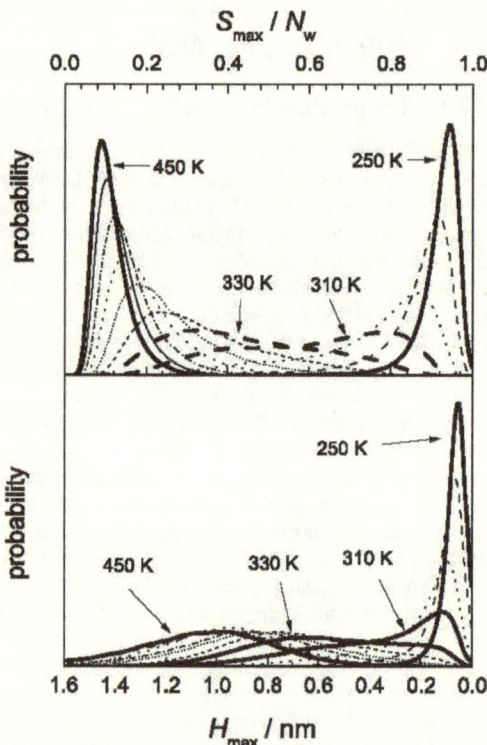

Figure 4.1: *Probability distribution of the fraction S_{max}/N_w of water molecules in the largest cluster (upper panel) and probability distribution of the distance H_{max} between the center of mass of the largest water cluster and that of the peptide (lower panel) at various temperatures in the hydration shell of hIAPP. The distributions for some temperatures are emphasized by thicker lines.*

The temperature dependence of the mean size S_{mean} of water clusters normalized by the number of water molecules in the hydration shells of hIAPP and hIAPP^S are shown in the lower panel of Figure 4.3, where S_{mean} is calculated by excluding the largest cluster. In accordance with the percolation theory, S_{mean} is a measure of the fluctuations of the size of clusters in the system studied, and since it shows a maximum at $T \approx (339 \pm 4)$ K

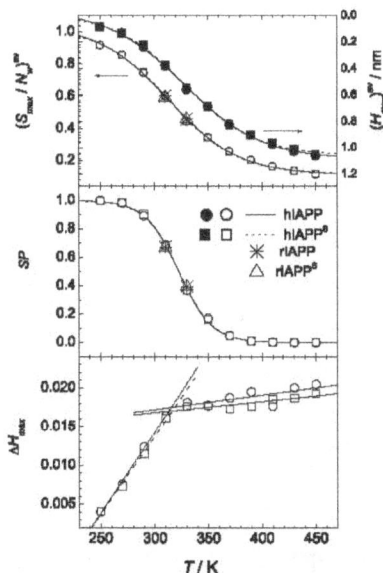

Figure 4.2: *Temperature dependencies of the average fraction* $(S_{max}/N_w)^{av}$ *of molecules in the largest cluster of hydration water (upper panel), of the spanning probability SP (middle panel), and of the standard deviation* ΔH_{max} *(lower panel) for water in the hydration shells of peptides (open circles and squares for hIAPP and hIAPP^S, respectively). The fits of the temperature dependencies SP(T) and* $(S_{max}/N_w)^{av}$ *to a sigmoidal function and* $\Delta H_{max}(T)$ *to a linear function are shown by lines. Data points for the two rat IAPP moieties are shown in the upper and middle panels for* $T = 310$ K *and* 330 K.

for both moieties, the strongest rearrangement of water clusters in the hydration shell occurs at about 339 K. S_{mean} diverges at the percolation transition in an infinite system and passes through the maximum, approaching the percolation transition in a finite system. Therefore, it is possible to conclude that the percolation transition of water in the hydration shells of hIAPP and hIAPP^S occurs at temperatures less than 339 K.

As mentioned in Section 1.4, one of the tedious aspects of applying the percolation theory is finding properties that can help define the percolation transition. One of them is H_{max}, the distance between the center of mass of the largest water cluster and the center of mass of a biomolecule, and it defines the spanning character of the largest cluster, i.e., its ability to envelope the biomolecule homogeneously. In the limit $S_{max}/N_w \rightarrow 1$, when all molecules in the hydration shell belong to the largest cluster being most homogeneously distributed at the surface, $H_{max} \rightarrow 0$ and the ability of the water network to envelope a biomolecule homogeneously is maximal. If the largest cluster covers only a small part of the biomolecules, H_{max} differs essentially from zero. The probability distributions $P(H_{max})$ for the largest water cluster in the hydration shell of hIAPP at various temperatures are shown in the lower panel of Figure 4.1. The data relative to the hIAPP^S distributions are quite similar (data not shown). In most configurations* at 250 K, H_{max} is less than just 0.1 nm, which means that the largest water cluster envelopes the peptide homogeneously at low temperatures. Upon heating, the probability distribution $P(H_{max})$ shifts rapidly to larger values of H_{max} and becomes much wider.

The temperature dependence of the average value $(H_{max})^{av}$ for the largest water cluster is shown in the upper panel of Figure 4.2 (NB, on the right axis), where one can

---

*The term configuration indicates that both the peptide and hydration shell are taken into account, as opposed to conformation, which refers to atom distributions that can be obtained without breaking bonds.

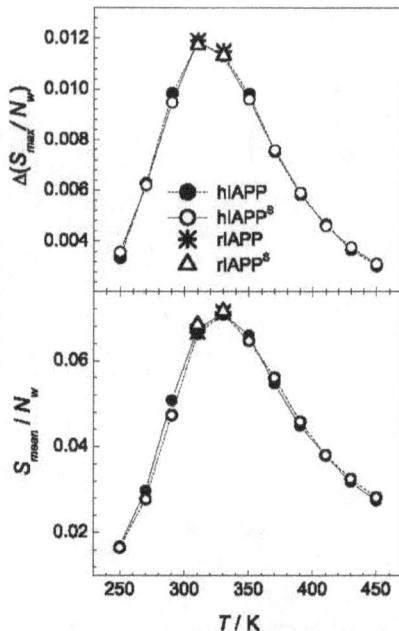

Figure 4.3: *Temperature dependencies of the standard deviation $\Delta(S_{max}/N_w)$ of the fraction of molecules in the largest cluster (upper panel) and of the mean size S_{mean} of clusters (lower panel) in the hydration shells of the IAPP variants, where the mean is calculated excluding the largest cluster.*

see it closely follows the temperature dependence of the percolation probability, i.e., the average fraction $(S_{max}/N_w)^{av}$ of water molecules in the largest cluster. The dependence of $(H_{max})^{av}$ on $(S_{max}/N_w)^{av}$ is shown in Figure 4.4, which can be perfectly fitted by the empirical equation:

$$(H_{max})^{av} = R \cdot (B - \beta (S_{max}/N_w)^{av}), \tag{4.1}$$

where $R = (1.48 \pm 0.02)$ nm, $\beta = 0.58 \pm 0.02$, and $B = 1.00$ for both hIAPP and hIAPP^S. Thus, $(H_{max})^{av}$ is equal to zero when all water molecules in the hydration shell are in one cluster. In the limit of very high temperatures, when all hydration water molecules have no H-bonds within hydration shell, the largest cluster includes a single water molecule and the value $(H_{max})^{av}$ is about 1.47 nm, exceeding slightly the radius of gyration of peptides considered (see below).

The temperature dependence of the width ΔH_{max} of the probability distribution $P(H_{max})$ for the largest water cluster is shown in the lower panel of Figure 4.2, where two distinct temperature regimes can be seen. Upon heating, ΔH_{max} increases strongly at $T < 320$ K and increases to a lesser extent at $T > 320$ K. The linear fits of these two regimes are shown by lines in the lower panel of Figure 4.2. The crossover between two regimes occurs at about 315 K and 316 K for hIAPP and hIAPP^S, respectively.

The probability distribution, n_S, of the water cluster size in the hydration shell, S, can also estimate the percolation transition, but this distribution is strongly affected by the finite system size. In fact, it never obeys the power law of $n_S \sim S^{-2.05}$ as expected at the percolation transition in 2D systems and should therefore not be used for the location of the percolation transition in finite systems, yet it still compares relatively well with

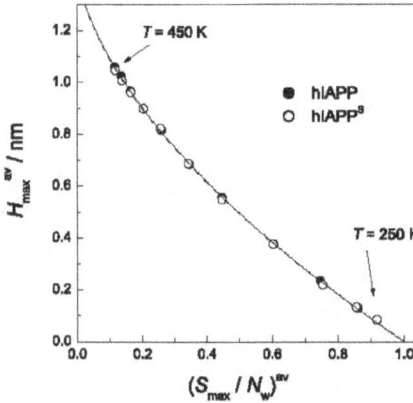

Figure 4.4: *Dependence of $(H_{max})^{av}$ on the average fraction $(S_{max}/N_w)^{av}$ of water molecules in the largest cluster. The fits to the Eq. (4.1) are shown by lines.*

the other methods utilized to determine the percolation transition (data not shown). It is therefore necessary to find an alternative method; in fact, the structure of the largest water clusters can be characterized by the fractal dimension, d_f, that describes how the mass distribution $m(r)$ of the largest cluster scales with distance r:

$$m(r) \sim r^{d_f}. \qquad (4.2)$$

The mass distribution is similar to a radial distribution function and was obtained by calculating the oxygen-oxygen distance between H_2O_i and each of the remaining $H_2O_{j\neq i}$ molecules of the largest cluster and summing up the oxygen-oxygen distances. The same procedure is carried out for each of the water molecules that are H-bonded. This distribution is also affected by the finite size of the system, as the number of water molecules in the largest cluster cannot be infinite, thus the mass distribution for the finite system diverges from the infinite one. The values of r for which the mass distribution behaves like an infinite system are kept, and those beyond it are discarded. The values d_f were obtained from the fits of the distribution $m(r)$ to Eq. (4.2) within the range $r<1.8$ nm; in fact, in this range, the mass of the largest cluster increases with r, whereas it decreases at $r>1.8$ nm as a result of the finite size of the cluster. The temperature dependence of the values d_f obtained from the fits for hIAPP and hIAPP^S are shown in Figure 4.5, where it can be seen that d_f achieves the largest value ≈ 2.20 at 250 K, which indicates that the hydration water is not strictly 2D, but rather a quasi-2D system.[*] A quite similar value ($d_f = 2.22$) was obtained in the hydration water shell studies at the surface of Snase.[107] The fractal dimension of percolating cluster is equal to 1.896 at the percolation threshold of 2D systems and d_f of the largest water cluster achieves this value at about 325 K, as seen in Figure 4.5. Taking into account that the fractal dimension of the whole shell of hydration water is about 2.2, it is reasonable to define the normalized fractal dimension $d_f^* = d_f(2/2.20)$,[†] as shown on the right axis in Figure 4.5, where such defined fractal dimension achieves the value 1.896 at ≈ 300 K.

---

[*]Owing to the height of the water shell, the hydration water is no longer ideally two-dimensional, but rather quasi-2D.

[†]This normalization is called upon to get an estimate of a 2D system from data obtained from a quasi-2D system.

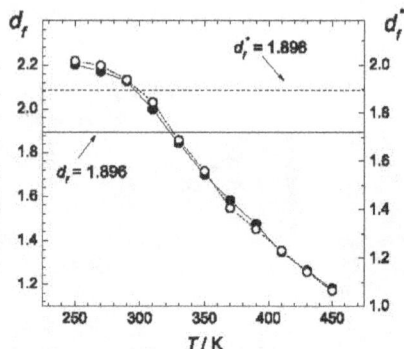

Figure 4.5: *Temperature dependence of the fractal dimension d_f of the largest water cluster in the hydration shells of hIAPP (solid circles) and hIAPP^S (open circles). The right scale shows d_f^* normalized by the deviation of the fractal dimension of the whole hydration shell from 2: $d_f^* = d_f \frac{2}{2.20}$. The horizontal line corresponds to the value d_f^* expected at the percolation threshold in two-dimensional systems.*

The arrangement of water molecules in the largest cluster can be characterized by the oxygen-oxygen pair correlation function, g_{OO}, calculated for the members of the largest water cluster, and can be seen for hIAPP at various temperatures in the upper panel of Figure 4.6, where these functions for hIAPP^S are quite similar, and thus not shown. At low temperatures, g_{OO} indicates tetrahedral-like arrangement of the water molecules in the largest cluster in the hydration shell; in fact, the second peak of g_{OO} is located at about 0.46 nm, close to the peak at 0.44 nm in bulk liquid water, as shown in the lower panel of Figure 4.6. Upon heating, this peak moves towards larger distances and is located at $r \approx 0.53$ nm at 450 K, close to the location of the peak of g_{OO} at $r \approx 0.55$ nm for hydration water near smooth hydrophilic surfaces and corresponds to the linear arrangement of oxygens of three water molecules. Therefore, the tetrahedral arrangement dominates in the structure of the largest clusters at $T \leq 330$ K, when most of these clusters are spanning. At higher temperatures, when most of the largest clusters are non-spanning, a linear-like arrangement dominates in their structure.

4.2.2   Effect of the Spanning Water Network on Peptide Properties

Various clustering properties of hydration water at the surfaces of hIAPP and hIAPP^S indicate that the thermal break of hydration water network occurs in the temperature interval between ≈ 300 K and ≈ 350 K. To clarify a possible effect of this break on the conformational properties of peptides, the temperature dependencies of the solvent accessible surface area, SASA, and the radius of gyration, R_g, have been analyzed with respect to the temperature dependence of the spanning probability, as can be seen in Figure 4.7. SASA of both hIAPP and hIAPP^S does not change notably at low temperatures from 250 K to 350 K (Figure 4.7, upper panel), whereas at higher temperatures the peptides show a sharp increase in SASA upon heating. A similar characteristic of the temperature dependence is observed also for the radius of gyration R_g of peptides (Figure 4.7, lower panel). Therefore, the structural properties of both hIAPP moieties undergo some qualitative changes, when the probability of observing water clusters including most of the molecules in the hydration shell becomes lower than approximately 0.2 (see lines in Figure 4.7, showing the temperature dependence of the spanning probability SP).

The stability of the secondary structure can be characterized by the width, $\Delta(n_H^{pp})$, of

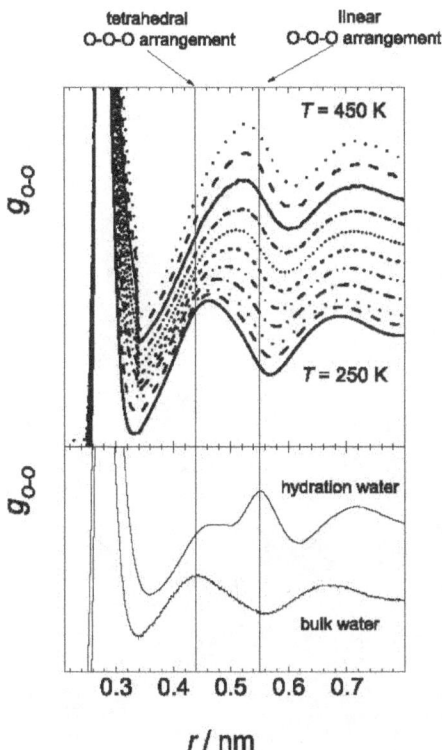

Figure 4.6: *Oxygen-oxygen pair correlation function g_{OO} calculated for the members of the largest water cluster in the hydration shell of hIAPP at various temperatures (upper panel, temperature increases from bottom to top). Oxygen-oxygen pair correlation function g_{OO} for bulk water and water in the hydration shell near smooth surface at $T=300$ K (lower panel). The vertical lines indicate the oxygen-oxygen distances, corresponding to the tetrahedral and linear arrangements of three oxygen atoms.*

the probability distribution, $P(n_H^{pp})$, of the total number of intrapeptide H-bonds, n_H^{pp}, as defined in Section 2.4.2. The value of $\Delta(n_H^{pp})$ reflects the ability of the peptide to undergo conformational changes accompanied by the fluctuations of the number of intrapeptide H-bonds: larger $\Delta(n_H^{pp})$ values indicate more conformational flexibility of peptides. The temperature dependence of $\Delta(n_H^{pp})$ for hIAPP and hIAPP^S is shown in Figure 4.8 together with the temperature dependence of the percolation probability $(S_{max}/N_w)^{av}$, where it is evidenced that a weaker H-bonded network of hydration water, i.e., a smaller fraction of water molecules in the largest cluster, corresponds to stronger fluctuations of n_H^{pp}.

A closer look into the secondary structure of the peptide as defined in Section 2.4.1, shows that, even though both the helical content, as seen in Figure 4.9b, and the H-bonds in Figure 4.9a decrease with temperature, they do not follow exactly the same trend. In other words, the H-bonds for both moieties decrease quasi-monotonically upon heating, with large scattering of the data as a result of the relatively problematic sampling for the ALL3 set of data, while the helical content is more or less constant below 320 K and decreases to another more or less constant value above 320 K.* The data relative to the

---

*In fact, the ALPH data have been fitted to a line with zero slope in a poster presented in Jülich, found in the Appendix B, Figure B.1, but this is mostly speculative, as there is plenty of scattering.

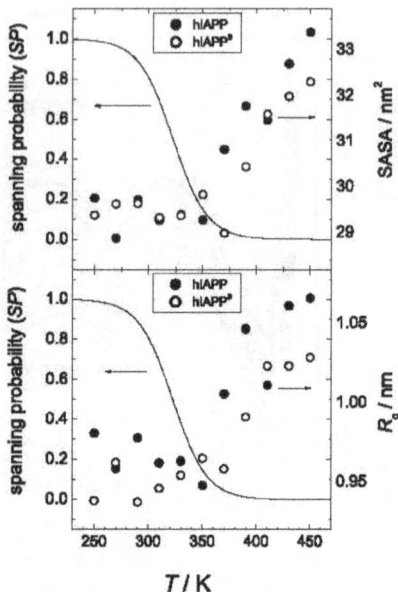

Figure 4.7: *Temperature dependence of the radius of gyration R_g (lower panel) and SASA (upper panel) of both hIAPP moieties (symbols). Temperature dependence of the spanning probability of H-bonded network in the hydration shell of both hIAPP moieties (lines). Both R_g and SASA start to increase upon heating, when a spanning network is almost broken.*

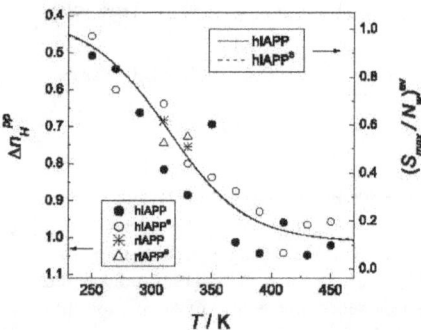

Figure 4.8: *Temperature dependence of the width $\Delta(n_H^{pp})$ of the probability distribution $P(n_H^{pp})$ of the number of intrapeptide H-bonds n_H^{pp} in hIAPP, $hIAPP^S$, rIAPP, and $rIAPP^S$ (symbols), and percolation probability $(S_{max}/N_w)^{av}$ (lines). NB, the values of $\Delta(n_H^{pp})$ decrease along the ordinate.*

secondary structure in Figure 4.9b does indeed show a large scattering and large uncertainty for certain runs, but better accuracy through longer simulation runs should give a more quantitative result. Another characteritic of the secondary structure that is strongly influenced by the percolation transition is the cooperative "condensation," as can be seen in Figures 4.9c and 4.9d. The probability n_s of finding S consecutive secondary structure elements in the same conformation in an uncorrelated distribution is[111]

$$n_s = (1-p)^2 p^S, \tag{4.3}$$

where p is the probability of finding the secondary structure of the studied biopolymer. The probability used in Eq. (4.3) is the mean value of helical content, p_{SS} in Figure 4.9b, obtained for 400 ns of concatenated data (i.e., the ALL3 trajectories, as described in Section 3.2) for the lower temperatures, and 200 ns for the higher temperatures (i.e., ALPH trajectories). The helical content at 310 K, as seen in Figure 4.9c, for the reduced moiety (right panel) deviates from the theoretical value obtained for an infinite chain (dashed line). As will be discussed in Chapter 5, the reduced moiety of IAPP seems to present more helical content, compared to the oxidized moiety. Moreover, it seems that the introduction of the natural disulfide bond, present between C2 and C7, also disrupts the cooperativity of the helices (left panel). At 330 K, the cooperativity observed at lower temperatures for the reduced moiety is close to random, as can be seen with the overlap with the the dashed line. The temperature effect seems to allow the oxidized peptide to explore even the longer helices, as can be seen in the left panel of Figure 4.9d.

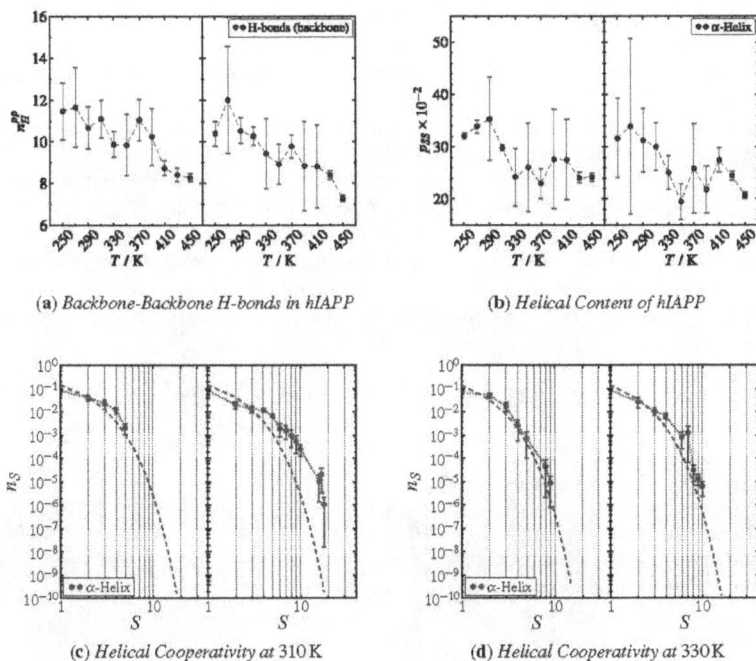

(a) Backbone-Backbone H-bonds in hIAPP

(b) Helical Content of hIAPP

(c) Helical Cooperativity at 310 K

(d) Helical Cooperativity at 330 K

Figure 4.9: Temperature dependence of (a) intrapeptide backbone-backbone H-bonds, n_H^{pp}, and (b) helical content percentage, p_{SS}, of oxidized (left panel) and reduced (right panel) hIAPP. The helical content is defined by the number of residues presenting the dihedral angles ϕ and ψ defined in Section 2.4.1. Probability n_S to find S successive residues with helical conformation at (c) 310 K and (d) 330 K. The dashed line shows n_S for a random distribution in an infinite chain with the same content probability p_{SS}, plotted in (b), for residues with analogous structure as expected by Eq. (4.3). The error bars indicate the estimated error calculated by g_analyze.

69

4.2.3 Effect of Peptide Structure on the Spanning Network of Hydration Water

The presence of a disulfide bond between C2 and C7 residues of hIAPP in the polypeptide does not affect noticeably the clustering of water molecules in the hydration shell of hIAPP, as can be seen from Figures 4.2 to 4.5 on pages 63–66. Although the amino acid content of rIAPP differs from that of hIAPP by about 16%, this modification has a negligible effect on the clustering of water molecules in the hydration shell; in fact, several parameters characterizing water clustering are compared for both moieties of hIAPP and rIAPP in Figures 4.2 and 4.3, allowing the conclusion that the thermal stability of the hydration water network on the surfaces of the studied hIAPP and rIAPP peptides is almost the same.

As will be discussed in detail in Chapter 5, the studied polypeptides do not show some well-defined secondary structure, so in order to study the effect of the peptide conformation on the clustering properties of water in its hydration shell, different conformations of hIAPP exhibiting essentially different structural properties were studied at 250 K. At such low temperature, the peptide conformation does not change noticeably during simulation runs of 50 ns, without being "frozen" (see discussion on the SPC/E water model density in Section 2.4.3.1, Figure 2.9). In addition to the main simulation run, two runs with less ordered initial peptide conformations, i.e., CL02 and CL03 seen in Section 3.2, characterized by a smaller number of intrapeptide H-bonds, n_H^{pp}, were simulated at 250 K. The obtained spanning probability, SP, and the percolation probability, $(S_{max}/N_w)^{av}$, for three quite different conformations of hIAPP are shown in Figure 4.10 as a function of n_H^{pp}, where, despite the strong scatter of the data, a clear correlation between two measures of the strength of the percolating network of hydration water and the number of intrapeptide H-bonds can be seen: the water network is stronger for conformations with larger values of n_H^{pp}.

Figure 4.10: *Dependence of the spanning probability (squares) and the percolation probability $(S_{max}/N_w)^{av}$ (circles) on the number of intrapeptide H-bonds n_H^{pp} in the main simulation run (closed symbols) and in two additional runs with lower values of n_H^{pp} (open symbols) at 250 K.*

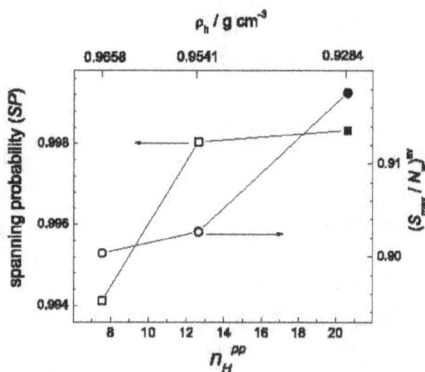

The change of the number of intrapeptide H-bonds, n_H^{pp}, affects the water-peptide interaction and thus the connectivity of water molecules in the hydration shell. Obviously, the decrease of n_H^{pp} causes an increase of the number of water-peptide H-bonds, and as a result, the surface of a peptide becomes more hydrophilic, causing the hydration water density to increase. Indeed, the density of hydration water, ρ_h, has been found to increase with decreasing n_H^{pp},[112] which should make the H-bonded water network stronger. However, the formation of direct water-peptide H-bonds has also an opposite effect, as it hinders the ability of water molecules to form water-water H-bonds; in fact, the weakening of the water network with decreasing n_H^{pp}, as seen in Figure 4.10, shows that the increase of density is not able to compensate for the damage of the H-bonded water network caused by formation of water-peptide H-bonds.

The change of the spanning probability upon varying n_H^{pp} is very small as it can be studied at very low temperature only, when SP is close to 1, where an estimation of the effect of n_H^{pp} on the temperature of the midpoint of the percolation can be carried out by shifting the fitting sigmoidal function $SP(T)$ (middle panel in Figure 4.2) in the temperature range in order to get the change of SP, which occurs upon decreasing n_H^{pp} at 250 K (Figure 4.10). Such estimation shows that the midpoint of the percolation transition shifts by just 5 K to lower temperatures, when n_H^{pp} decreases from ≈ 20 to ≈ 8.

4.3 Conclusions

Analysis of the effect of temperature on the clustering of hydration water at the surfaces of IAPP variants evidences a thermally induced break of the spanning H-bonded network of hydration water into an ensemble of small, i.e., finite, clusters via a percolation transition. Various properties describing clustering and percolation show a strong qualitative change of the connectivity of H-bonds within the hydration shell of IAPP in the temperature interval from 300 K to 330 K, and as shown in Figure 4.2, the midpoint of the percolation transition occurs at approximately 320 K. Close to this temperature, the fluctuations of the size of the H-bonded water clusters in the hydration shell are maximal, as seen in Figure 4.3. The fractal dimension of the largest cluster (Figure 4.5) and the average number of water-water H-bonds formed by one molecule (data not shown) achieves values corresponding to the 2D percolation threshold just above 300 K; in fact, the ability of the H-bonded water network to homogeneously cover a peptide changes qualitatively at about 310 K, as shown in Figure 4.2.

The analysis presented in this chapter indicates that the break of the spanning water network correlates with the essential structural changes of IAPP; in fact, when the peptide is enveloped by a spanning water network, its compactness, characterized by the radius of gyration R_g and by the solvent accessible surface area SASA, does not change with temperature, while in the absence of this network, both R_g and SASA increase upon heating. This trend correlates with the helical content; in fact, despite the scattering of the data, the helicity of the peptide at lower temperatures is more or less constant, while at higher temperatures, it drops to lower temperature-independent values. Moreover, the conformational flexibility of a peptide, characterized by the fluctuations of the number of intrapeptide H-bonds, shows negative (or anti-) correlation with the "strength" of the network of hydration water, characterized by the fraction of water molecules in this network.* This is in accord with the studies of the spanning water network on the con-

---

*Although it may be highly speculative, the trends seen for the concatenation of independent runs in Figure 3.6b on page 55 also show better sampling starting from 290 K and, in particular, above 310 K, as in Figure 3.7b on page 57. It is obvious that the sampling is better at higher temperatures, but it would be interesting to investigate whether this rigidity of the peptide in exploring R_g is caused by the water shell or not; a few runs in vacuo at lower temperatures could possibly answer that question.

formational flexibility of ELP;[41] in fact, at low temperatures, it exhibits long-lived rigid conformations controlled by the intrapeptide H-bonds between distant residues, while, when the spanning network of the hydration water breaks upon heating, the flexibility of ELP increases drastically, exhibiting structural properties typical of a random chain.

The variation of the chemical structure of the peptide as well as the variation of its conformation, controlled by the total number of intrapeptide H-bonds, practically does not affect the thermal stability of the hydration water network, which holds true also for the capping of hIAPP variants. The temperature shift of various parameters describing the connectivity of H-bonds in the hydration shell due to these changes does not exceed a few degrees, as seen in Figures 4.2 to 4.5 on pages 63–66. Such universality of the temperature range, where the spanning network of hydration water breaks, is even more evident when the thermal stability of the hydration of water at the surface of several different biomolecules are compared (data not shown), in which the size and the degree of the hydrophilicity/hydrophobicity vary strongly in the set of these molecules. The thermal stability of the spanning network of hydration water is, nevertheless, surprisingly similar for all the studied molecules.

The temperature interval, where the strongest changes of the connectivity of H-bonds in the hydration water shell occur, is between 300 K and 330 K; temperature interval in which the activity of living organisms is maximal.[106] The body temperature of the warm-blooded organisms, i.e., the only organisms that can keep their body temperature almost constant, is between 310 K and 320 K. As biological functions are possible in the presence of hydration water only, the occurrence of the maximal biological activity at temperatures where the connectivity of H-bonds in the hydration shell undergoes strong changes does not seem to be accidental, although it is still not clear which particular property of hydration water promotes strong biological activity in this temperature range.

The changes in the conformation and aggregation behavior of many biomolecules are the strongest between 300 K and 330 K; in particular, the experimentally measured lag time of hIAPP aggregation drops drastically at about 320 K.[10] Besides, many biomolecules undergo conformational transition and aggregation in this temperature interval, possibly related to the qualitative changes of the hydration water network in the same temperature interval. As has been shown by Oleinikova et al., the thermal break of the H-bonded water network causes sharp changes of the components of the specific heat of hydration water, related to water-water interactions within the hydration shell and to the interactions between hydration water and surrounding "bulk-like" water.[42] These changes enhance the connectivity between hydration and bulk water; as a result, the surface of a biomolecule is effectively more hydrophobic, which may thus provoke denaturation and aggregation of biomolecules.

Chapter 5

Comparing
hIAPP and rIAPP in Liquid Water

A mystery that has been keeping many scientists busy is how completely different amino acid sequences can form aggregates in amyloidoses. The data presented in this study definitely cannot answer that question alone, although there are a few findings that can shed a little light on the behavior of monomeric islet amyloid polypeptide in liquid water, since the human homologue does aggregate and is present in more than 95% of the type II diabetes patients, whereas the rat/mouse homologue does not, even though the primary structure differs by only 16%. As demonstrated in Chapter 3, even gathering the data from the trajectories was a bit problematic since the peptide lacks a well-defined conformation. In order to sample sufficient conformations, 500 ns trajectories were calculated starting from a completely random conformation. The temperature dependencies obtained in Chapter 4 show that the peptide undergoes a conformational transition when heated to 330 K, also observed experimentally by Kayed et al. since the lag time in aggregation drastically dropped around 320 K.[10] Therefore, the choice of temperatures for the sampling of conformations was limited to 310 K and 330 K, which also happen to be physiologically relevant temperatures.

5.1 Conformational Changes of Oxidized hIAPP at 330 K

In the investigation of possible initial conformations of IAPP for the MD simulations, it was observed that secondary structure had a noticeable effect on reducing R_g by 8% to 15% in one of the runs of oxidized hIAPP at 350 K (HoRUN4 in Table 3.1), which seemed an outlier as a result of the small R_g, as seen in Figure 3.2b on page 45 of Section 3.1.1.2. It was found that there was a low, yet stable, β-structural content of (5.1 ± 0.5)%, making its structure more compact. In fact, further investigation of the effect of β-structures on oxidized hIAPP at 330 K was carried out and is discussed in this chapter.

A very striking characteristic of hIAPP is the flexibility of the oxidized moiety shown in reaching compact conformations, as opposed to rIAPP, which appears more rigid, as the greater helical content might suggest. The presence of the disulfide bond seems to play a role in the initiation of the collapsing of the peptide in virtue of the stability of the threonine/disulfide interaction. The collapsed state is characterized by β-bridges and β-ladders, short tyrosine/phenylalanine distance, and diminished hydrophobic SASA. Collapsed states are not reached by the reduced hIAPP, perhaps as a result of the absence of the disulfide bond and the presence of transient α-helices in the C-terminal region. All the forms of rIAPP studied, with the exception of the oxidized moiety at 310 K, seem too rigid to fold into conformations characterized by short r_{eted}, which seems to be necessary for collapsing into the compact monomeric state. This rigidity seems to be caused by helicity that includes P28; in fact, as seen in Figure 5.13c, there is no helical content, i.e., α and 3_{10} helices, when r_{eted} is minimum, as seen in Figure 5.2a (top panel, in blue).

5.1.1   Conformational Properties of IAPP—R_g, r_{eted}, and SASA

The conformational properties for each moiety of hIAPP—R_g, L_{max}, and SASA—in-crease with temperature for the cysteine moiety, as does the standard deviation of the mean, indicating a greater flexibility of the peptide with increasing temperature. Mean values of the conformational properties r_{eted} and R_g, with the estimated error, are shown in Tables 5.1 and 5.2, but as will be explained later in this section, the oxidized moiety collapses into a compact state and thus only the first 350 ns of the trajectory can be used to estimate average values of conformational properties before the peptide collapses. In Table 5.2, the subscript "$_C$" refers to the collapsed state of the oxidized moiety, corre-sponding to the last 150 ns, while the subscript "$_E$" refers to the first 350 ns, excluding therefore the collapsed conformation. The same subdivision of the data is also shown for the reduced moiety, as a reference; in fact, the reduced hIAPP moiety at 330 K aver-ages out to ≈ 0.98 nm, whether the data are taken from the first portion of the trajectory, "E," the last, "C," or the entire trajectory. Moreover, the reduced moiety increases with temperature from (0.948 ± 0.004) nm to (0.984 ± 0.012) nm, as seen in Tables 5.1 and 5.2. At 310 K, the cystine moiety appears more flexible and able to explore more states than those visited by the cysteine moiety at the same temperature, as can be seen by the fluctuation of R_g and r_{eted} in Table 5.1. Fluctuations in these conformational properties can be seen in the time-dependent plots of R_g and r_{eted} shown in Figures 5.1 and 5.2, respectively. The most striking feature is the compactness of the peptide, defined by the low values of R_g, observed in the upper panel of Figure 5.1c from 350 ns on, i.e., trajectory "C." Moreover, such values are much less than the theoretical value obtained for a random-flight chain, depicted in cyan. Hence at 330 K, the oxidized hIAPP moiety reaches a state smaller than that of an unperturbed random coil, by means of interac-tions between aromatic residues and backbone-backbone intrapeptide H-bonds, as will be demonstrated in the following sections. This collapsed state of oxidized hIAPP at 330 K presents an R_g 9.0 % smaller than the mean value of R_g of the 350 ns before the collapse, 11.2 %* smaller than the reduced moiety at 330 K, and 9.5 %† smaller than the mean value of all the rIAPP runs at 330 K, results that agree with the higher degree of folding of monomeric hIAPP found by Soong et al. [113]

A cluster analysis on every tenth frame of the hIAPP cystine moiety simulation run at 330 K, by means of the GROMOS algorithm[91] using a cutoff of 0.3 nm on the backbone atom RMSD, shows that the family of compact conformations, including the one stabi-

---

*Errata corrige: in Ref. 44, it should read 11.2 % instead of 11.1 % (typo from 11.19 %).

†Errata corrige: in Ref. 44, it should read 9.5 % instead of 9.9 % (typo from 9.49 %).

lized by β-structures, is present 39 % of the time, i.e., at least 9 % more than what would be expected from 150 ns of the compact stabilized conformation. In fact, the second most populated family of structures of the first 350 ns, consisting of 11 % of the total number of conformations of the initial portion of the trajectory, contains the centroid conformation of the most populated family of structures of the full simulation. Therefore, a compact conformation is already present before the peptide collapses. Further investigation on how the peptide "folds" into a more compact conformation is possible through the analysis of these data and should shed light on the peptide "folding." The word folding is used loosely here, since the peptide is natively unstructured, and the pathway to the observed collapsed conformation is not intended to describe the native conformation.

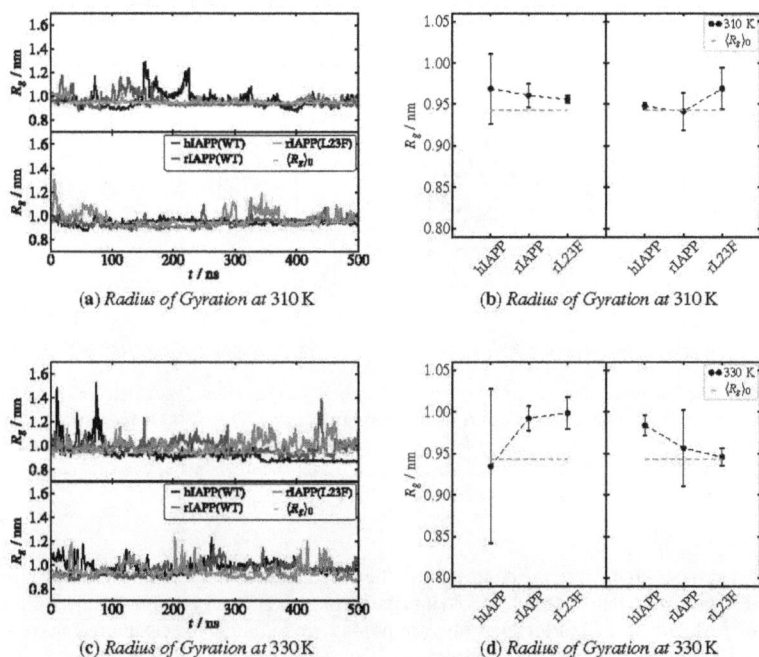

(a) *Radius of Gyration at* 310 K

(b) *Radius of Gyration at* 310 K

(c) *Radius of Gyration at* 330 K

(d) *Radius of Gyration at* 330 K

Figure 5.1: *The time-dependent data relative to oxidized (top panel) and reduced (lower panel) hIAPP at (a) 310 K and (c) 330 K, while their corresponding average values are shown in (b) and (d), with the oxidized and reduced moieties in the left and right panels, respectively. The mean values are found in Tables 5.1 and 5.2. The dashed cyan line indicates the values calculated for the random-flight chain seen in Table 2.3.*

One aspect that seems to be necessary for the ability of the peptide to form this collapsed state is the presence of the disulfide bond;[35] in fact, in the top panel of Figure 5.2c, one can see a short end-to-end distance, not only in the final 150 ns, but also at ≈200 ns, albeit short-lived. Hence, the oxidized moiety appears flexible enough to fold and collapse

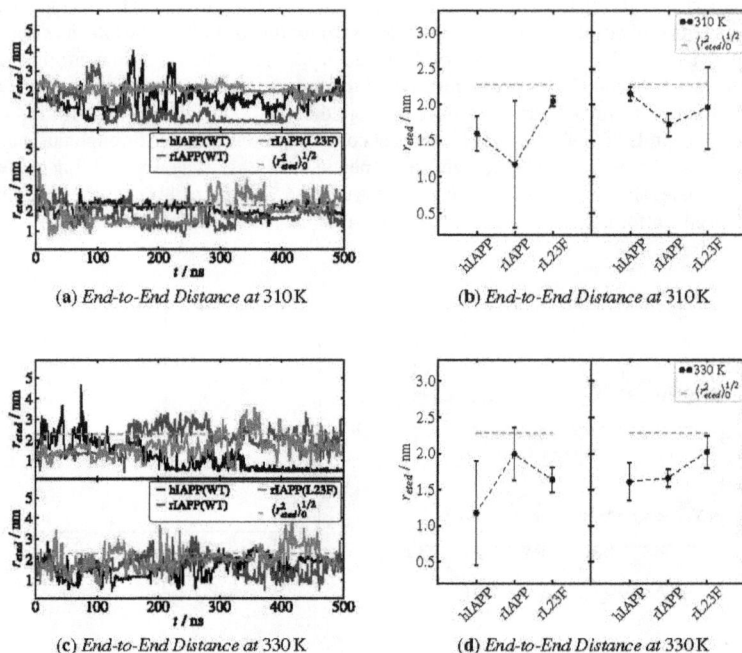

(a) *End-to-End Distance at* 310 K

(b) *End-to-End Distance at* 310 K

(c) *End-to-End Distance at* 330 K

(d) *End-to-End Distance at* 330 K

Figure 5.2: *The time-dependent data relative to oxidized (top panel) and reduced (lower panel) hIAPP at (a) 310 K and (c) 330 K, while their corresponding average values are shown in (b) and (d), with the oxidized and reduced moieties in the left and right panels, respectively. The mean values are found in Tables 5.1 and 5.2. The dashed cyan line indicates the values calculated for the random-flight chain seen in Table 2.3.*

into a compact state at temperatures above the percolation transition, i.e., 320 K, while rIAPP seems too rigid to be able to fold into more compact states, as also shown by Vaiana et al. [35] The reduced hIAPP moiety also seems too rigid to fold in the central region (residues 10–20), as a result of the high helical content in that region, as seen in Figure 5.13b; in fact, the mean r_{eted} measured for all the rIAPP runs at 330 K is between 1.64 nm and 2.0 nm and (1.6 ± 0.3) nm for the reduced hIAPP moiety. On the other hand, the cystine moiety reaches a very short r_{eted} of (0.54 ± 0.02) nm, confirmed by the H-bond between T36/C2 shown in Figure 5.19a, present only in this folded peptide. This can be explained by the interactions between residues along the chain and the disulfide bond of the cystine moiety, which do not occur in the cysteine moiety. [35] There are, in fact, side chain interactions with the disulfide region between T36 and both C2 and C7, albeit with an occurrence of 20 %–25 % each (data not shown). A low value of r_{eted} was also observed in oxidized rIAPP at 310 K (blue line in the top panel of Figure 5.2a), underlining the importance of the disulfide bond in the stabilization of short end-to-end distances, confirmed also by an

H-bond between N35/C7 (occurrence $>50\%$, data not shown), although it is not enough to allow the peptide to fold into a more compact conformation. As can be seen in the top panel of Figure 5.9c, there is no aromatic-aromatic interaction to stabilize this state, for it is the Y37/L23 distance that is short, not Y37/F15. In hIAPP, residue twenty-three is aromatic, i.e., phenylalanine, while in rIAPP, it is an aliphatic chain, i.e., leucine. Further observations on the aromatic-aromatic interactions are made in Section 5.2.1.

Table 5.1: *Conformational Properties r_{eted} and R_g at* 1 bar *and* 310 K

	$\langle r_{eted} \rangle$ / nm	$\langle R_g \rangle$ / nm		$\langle r_{eted} \rangle$ / nm	$\langle R_g \rangle$ / nm
Ox. hIAPP	1.6±0.2	0.97±0.04	Red. hIAPP	2.15±0.09	0.948±0.004
Ox. rIAPP	1.2±0.9	0.961±0.014	Red. rIAPP	1.72±0.16	0.94±0.02
Ox. rL23F	2.05±0.07	0.956±0.004	Red. rL23F	2.0±0.6	0.97±0.03

Table 5.2: *Conformational Properties r_{eted} and R_g at* 1 bar *and* 330 K

	$\langle r_{eted} \rangle$ / nm	$\langle R_g \rangle$ / nm		$\langle r_{eted} \rangle$ / nm	$\langle R_g \rangle$ / nm
Ox. hIAPP	1.2±0.7	0.93±0.09	Red. hIAPP	1.6±0.3	0.984±0.012
Ox. rIAPP	2.0±0.4	0.992±0.015	Red. rIAPP	1.66±0.12	0.96±0.05
Ox. rL23F	1.64±0.17	0.999±0.019	Red. rL23F	2.0±0.2	0.946±0.010
Ox. hIAPP$_E$	1.4±0.9	0.96±0.08	Red. hIAPP$_E$	1.48±0.12	0.986±0.015
Ox. hIAPP$_C$	**0.54±0.02**	**0.8738±0.0009**	Red. hIAPP$_C$	1.9±0.4	0.98±0.04

Investigating the final portion of the 330 K simulation run of oxidized hIAPP, the frequency distribution of SASA (Figure 5.4a, bottom) shows the reduction of SASA; in fact, the entire run clearly shows two states: a larger one, before the peptide folds into the compact state, and a smaller one, after it collapses. This can be seen not only for the total SASA (Figure 5.4a, in red), but in particular for the fraction of SASA relative to the hydrophobic residues (Figure 5.4a, in blue), hence hydrophobic interactions seem to push the peptide towards a more compact conformation. The overall trend seen in Figures 5.3 and 5.4 is that the hydrophilic fraction of SASA is less than the hydrophobic fraction; in fact, the hydrophilic distribution (in green) is always to the left of the hydrophobic one (in blue). Another characteristic seen in these SASA distributions is the fact that small hydrophobic SASA exposures are more frequent for hIAPP than for the rIAPP variants; in fact, oxidized hIAPP at both temperatures and reduced hIAPP at 330 K, seen in Figures 5.3a, 5.4a, and 5.4b, respectively, show a significant amount of conformations displaying hydrophobic SASA that is less than 15 nm^2, substantially lower than the initial value.* On the other hand, the hydrophilic SASA exposures are more or less equally distributed around the initial value. This could mean that hIAPP is more flexible than the rIAPP variants to be able to fold and "tuck under" the hydrophobic parts, with the cystine moiety being more flexible at lower temperature.

The correlation between R_g and SASA could possibly shed some light also on the shape of the peptide, i.e., if the peptide is in a spherical conformation, SASA correlates with R_g^2, while if it is more similar to an elongated ellipsoid, it correlates with R_g.[67] As

---

*The SASA of the initial α-helical conformation is 20.5 nm^2 and 13.0 nm^2 for the hydrophobic and hydrophylic solvent accessible surface areas, respectively.

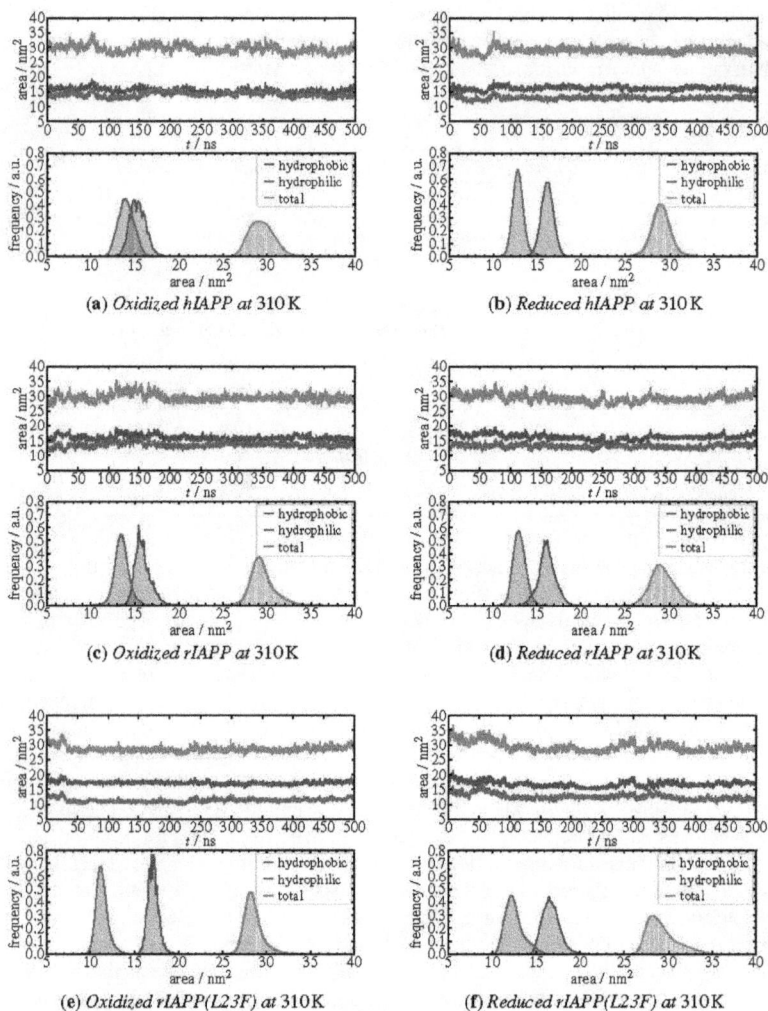

Figure 5.3: *Time dependence of the SASA (top) and frequency distribution of the data (bottom) at 310 K. Hydrophilic (in green), hydrophobic (in blue), and total (in red) SASA for (a) oxidized and (b) reduced hIAPP; (c) oxidized and (d) reduced rIAPP; (e) oxidized and (f) reduced rIAPP(L23F).*

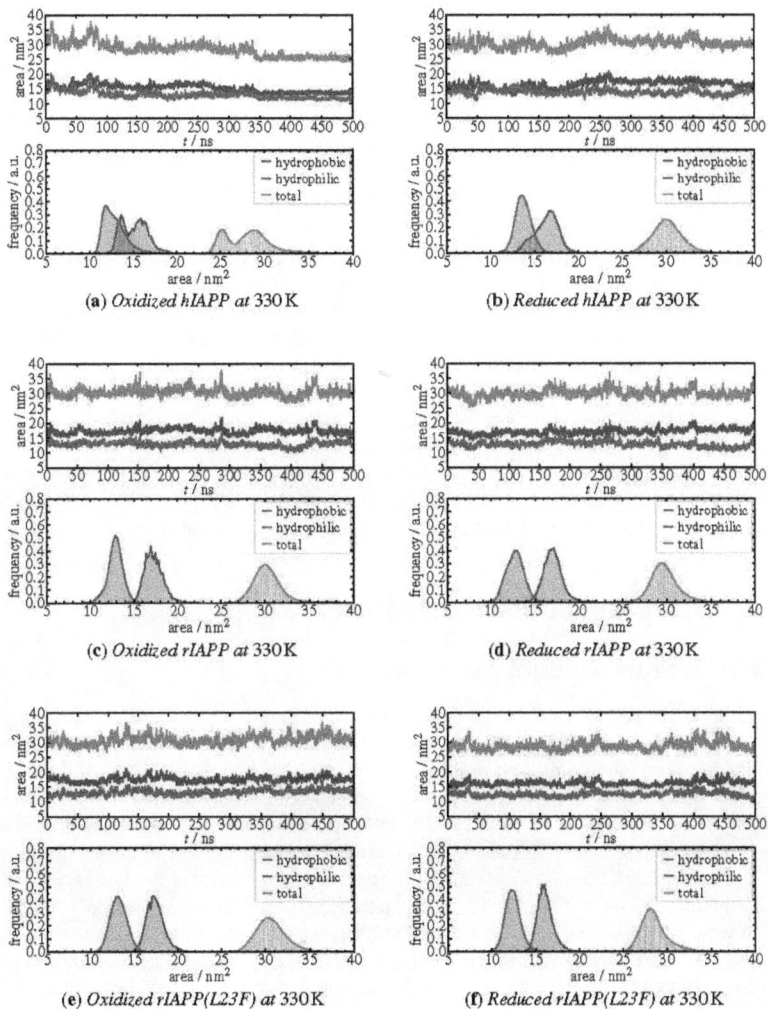

Figure 5.4: *Time dependence of the SASA (top) and frequency distribution of the data (bottom) at 330 K. Hydrophilic (in green), hydrophobic (in blue), and total (in red) SASA for (a) oxidized and (b) reduced hIAPP; (c) oxidized and (d) reduced rIAPP; (e) oxidized and (f) reduced rIAPP(L23F).*

seen in Figures 5.5 and 5.6, at the two temperatures studied, the correlation presents co-efficients r that are at least 0.60, with the exception of reduced rIAPP at 310 K, as seen in Figure 5.5d. There is no overall clear difference between the correlation coefficients r_{xy} and r_{x^2y}, if not in the case of the oxidized hIAPP; in fact, in Figure 5.6a, the correlation coefficient is at least 0.81, with it improving to 0.84 when the correlation between R_g and SASA is linear, i.e., r_{xy}. It is, therefore, not possible to discern between a spherical or cylindrical/elongated ellipsoid-like conformation, although there is, in most cases, a slight preference for the latter. One possible explanation is that the values of R_g are very close to 1 nm, as seen in Tables 5.1 and 5.2, so for these data this relationship between SASA and R_g is not accurate enough to distinguish between different powers of R_g.

A temperature dependence of the previously mentioned conformational properties for the cystine moiety cannot be established without a bias, as the peptide folds into a compact state by forming β-bridges. In fact, both moieties show a decrease in helical content with increasing temperature, while the number of H-bonds (n_{HB}) decreases with increasing temperature for the cysteine moiety and increases for the cystine moiety. An anti-correlation can be observed between n_{HB} and R_g at 330 K, where the correlation coefficient r_{xy} is -0.66 (see Figure 5.8a), compared to the remaining conformations in Figures 5.7 and 5.8, underlining the compactness of the peptide. If the data points are reduced by block averaging, the correlation coefficient increases in the following manner: -0.74, -0.77, and -0.82, when blocks are formed by averaging over 50 ps, 500 ps, and 5 ns, respectively. The correlation of the 100 mean values, obtained through block averages over 5 ns, is good, until the last data points, in which the R_g has reached its minimum, and the total number of H-bonds still increases (data not shown).

5.2   Compact hIAPP Conformation at 330 K

5.2.1   Aromatic-Aromatic Interactions

In solution, the peptide seems to be rather compact, and although essentially random coil in nature, there are some ordered structural elements that can give additional information with regards to the aggregation propensity of the peptide. Aromatic residues, such as tyrosine and phenylalanine, seem to play an important role in aggregation. In the mature fibers, Y37 is close to both F15 and F23,[27] while the hIAPP peptide in solution does not have an ordered structure.[10] Measuring the distance between the $C\alpha$'s of these residues at 310 K and 330 K, the average values of the time-dependent plots in Figures 5.9 and 5.10 fluctuate between 0.5 nm and 2.5 nm. This clearly shows that IAPP is more compact than what would be expected for a completely random peptide; in fact, a random walk model[30] of the polypeptide between residues Y37/F23 and Y37/F15 predicts distances of 3.0 nm and 4.0 nm, respectively.[27] As the peptide seems unstructured in solution, the Förster distances between the tyrosine and phenylalanine can be related to the radius of gyration. If not in the hIAPP cystine run at 330 K, there is no direct correlation between R_g and Y/F distances. Since aromatic residues that are farther than 1.8 nm apart cannot give reliable FRET results, a peptide in solution with the average distance greater than such value will be essentially unstructured. There seems to be no marked temperature dependence with this conformational property. Moreover, similar distances are found in rodent IAPP and since F23 is not present in rIAPP, interaction between aromatic

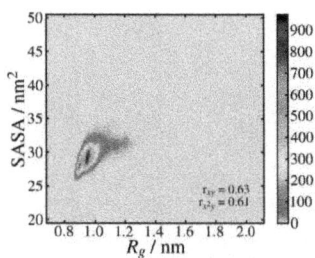

(a) *Oxidized hIAPP at* 310 K

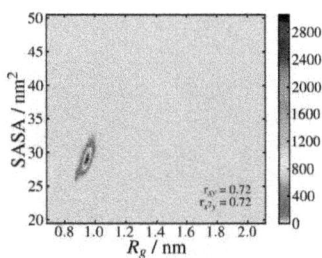

(b) *Reduced hIAPP at* 310 K

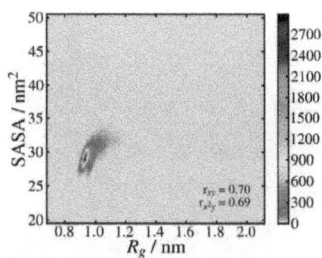

(c) *Oxidized rIAPP at* 310 K

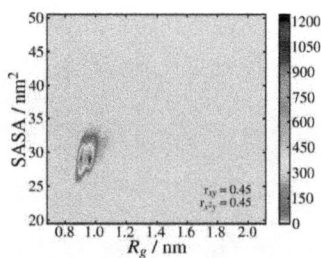

(d) *Reduced rIAPP at* 310 K

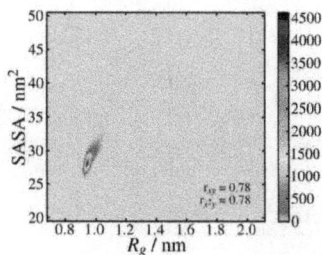

(e) *Oxidized rIAPP(L23F) at* 310 K

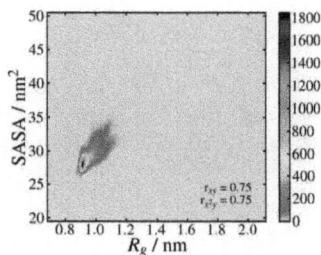

(f) *Reduced rIAPP(L23F) at* 310 K

Figure 5.5: *Radius of Gyration, R_g, and solvent accessible surface area, SASA, correlation at 310 K for (a) oxidized and (b) reduced hIAPP; (c) oxidized and (d) reduced rIAPP; (e) oxidized and (f) reduced rIAPP(L23F). The correlation coefficients, r_{xy} and r_{x^2y}, where the latter correlates the SASA against R_g^2, were calculated with* NumPy.

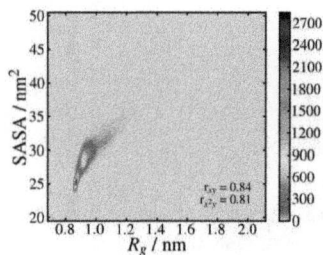

(a) *Oxidized hIAPP at* 330 K

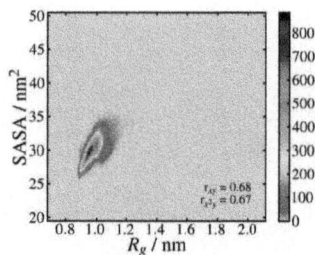

(b) *Reduced hIAPP at* 330 K

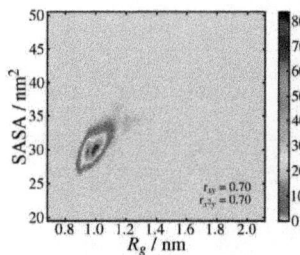

(c) *Oxidized rIAPP at* 330 K

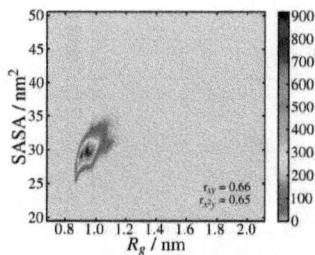

(d) *Reduced rIAPP at* 330 K

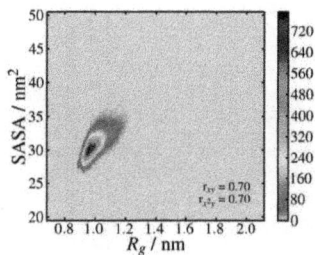

(e) *Oxidized rIAPP(L23F) at* 330 K

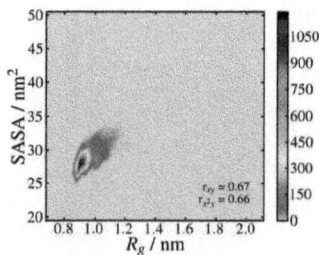

(f) *Reduced rIAPP(L23F) at* 330 K

Figure 5.6: *Radius of Gyration, R_g, and solvent accessible surface area, SASA, correlation at 330 K for (a) oxidized and (b) reduced hIAPP; (c) oxidized and (d) reduced rIAPP; (e) oxidized and (f) reduced rIAPP(L23F). The correlation coefficients, r_{xy} and r_{x^2y}, where the latter correlates the SASA against R_g^2, were calculated with* NumPy.

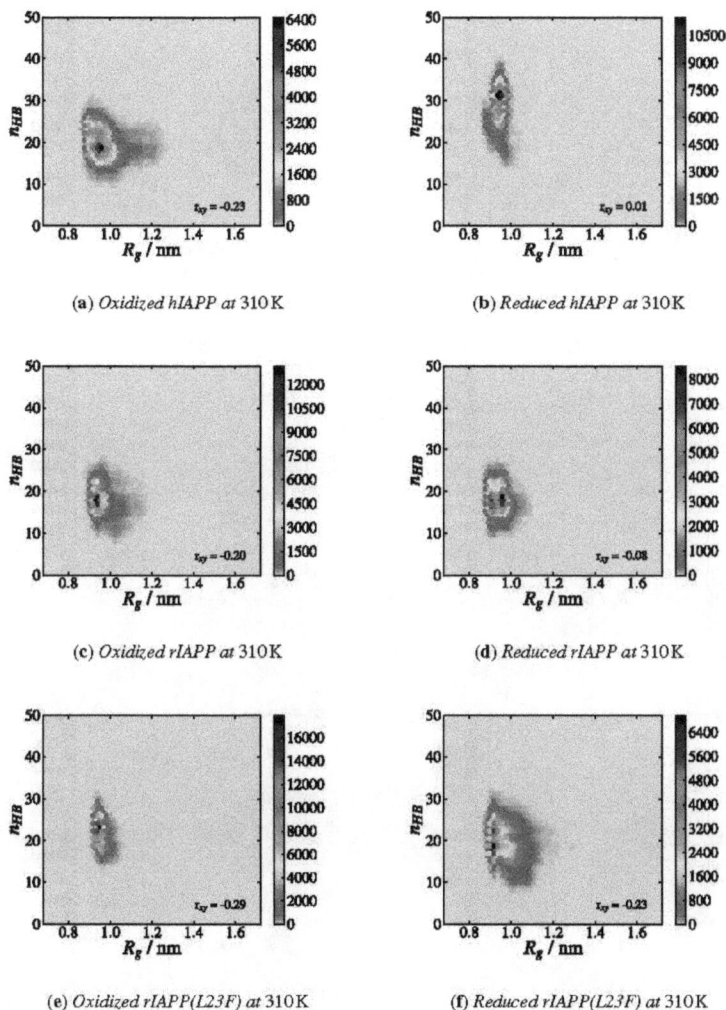

(a) *Oxidized hIAPP at* 310 K

(b) *Reduced hIAPP at* 310 K

(c) *Oxidized rIAPP at* 310 K

(d) *Reduced rIAPP at* 310 K

(e) *Oxidized rIAPP(L23F) at* 310 K

(f) *Reduced rIAPP(L23F) at* 310 K

Figure 5.7: *Radius of Gyration, R_g, and number of H-Bonds, n_{HB}, correlation at 310 K for (a) oxidized and (b) reduced hIAPP; (c) oxidized and (d) reduced rIAPP; (e) oxidized and (f) reduced rIAPP(L23F). The correlation coefficient, r_{xy}, was calculated with* NumPy.

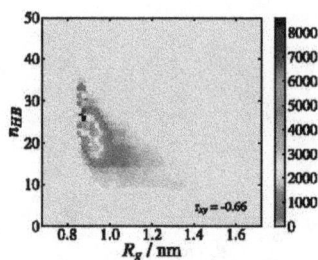

(a) *Oxidized hIAPP at* 330 K

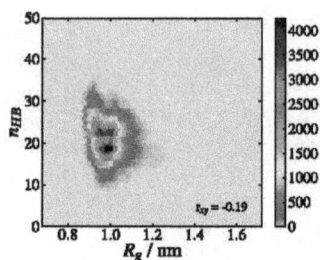

(b) *Reduced hIAPP at* 330 K

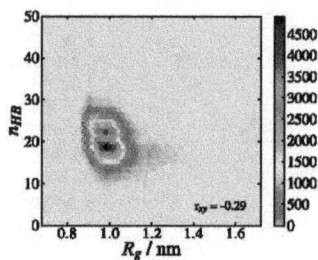

(c) *Oxidized rIAPP at* 330 K

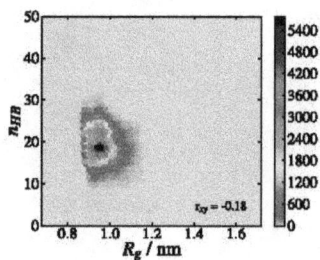

(d) *Reduced rIAPP at* 330 K

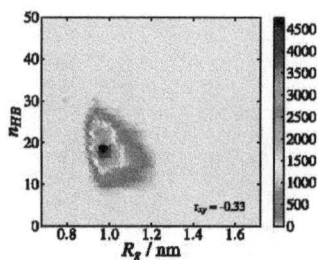

(e) *Oxidized rIAPP(L23F) at* 330 K

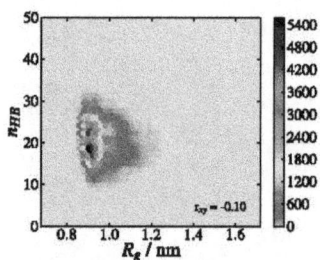

(f) *Reduced rIAPP(L23F) at* 330 K

Figure 5.8: *Radius of Gyration, R_g, and number of H-Bonds, n_{HB}, correlation at* 330 K *for (a) oxidized and (b) reduced hIAPP; (c) oxidized and (d) reduced rIAPP; (e) oxidized and (f) reduced rIAPP(L23F). The correlation coefficient, r_{xy}, was calculated with* NumPy.

residues can be excluded as the sole cause of stabilizing the single IAPP peptide conformation in solution. In fact, Marek et al. have shown that intramolecular or intermolecular
aromatic-aromatic or aromatic-hydrophobic interactions are not required for IAPP amyloid formation, by creating a triple mutant that does not contain aromatic residues, i.e.,
phenylalanine and tyrosine have been replaced by leucine. The aromatic residues are,
however, relevant for the kinetics and influence the topology of the deposits. [114] From the
data collected for the in silico mutated rIAPP sequence, in which leucine was mutated
into phenylalanine (L23F), found to form aggregates at small yields, [14] there do not seem
to be any significant differences in the Y37/F23|L23 and Y37/F15 distances between the
rIAPP, hIAPP, and rIAPP(L23F) moieties, as can be seen in the top panels of Figures 5.9
and 5.10. Through the data collected of the Y/F distances for oxidized hIAPP at 330 K,
it is possible to shed light on the mechanism of how the peptide folds into a compact
structure, as seen in Figure 5.4a. The correlation seen in Figure 5.6a, along with visual
comparison of the data, allow the compactness of the peptide to be described by either the
total SASA or R_g. While the peptide is compact, as seen by the small SASA from 100 ns
onwards (Figure 5.4a, top), these aromatic residues are still mobile and not buried within
the peptide surface. The Y37/F23 distance fluctuates greatly from 0.4 nm to 2.3 nm, until it reaches the value of 1.1 nm at $\approx$300 ns (Figure 5.10a top, blue), comparable to
1.26 nm, which Padrick and Miranker measured during the lag phase in hIAPP. [27] After
approximately 50 ns upon reaching this minimum, the Y37/F15 distance also reaches minimum $\approx$0.8 nm (Figure 5.10a top, green), presenting more fluctuations than the Y37/F23
distance. The Y37/F23 value averaged over the last 150 ns is (1.103 ± 0.002) nm and
Y37/F15 is (0.86 ± 0.10) nm, which correspond to a high content of backbone-backbone
H-bonds given by the β-structures (Figure 5.10a bottom, green). As can be seen in
Figure 5.10a, the short Y37/F15 distance is necessary, but not sufficient, to fold the
peptide; in fact, between 80 ns and 120 ns these Y/F distances are very short, but the
radius of gyration is $\approx$0.95 nm and not 0.87 nm as measured in the final 150 ns (top panel
in Figure 5.1c), the former comparing nicely to the value of 0.94 nm calculated for a
random-flight chain, as seen in Section 2.4.3. In that same time frame, between 80 ns and
120 ns, r_{eted} is also relatively close to the value obtained for a random-flight chain, and not
(0.54 ± 0.02) nm, as observed in the collapsed conformation (top panel in Figure 5.2c).

Of the secondary structure assigned by DSSPcont, the β-structures, which include
both β-bridges and β-ladders,* show a steady increase to $\approx$20 %, starting from $\approx$80 ns,
which corresponds to the minimum Y37/F15 distance (Figure 5.10a bottom, green). By
comparison, the reduced moiety also presents very short Y37/F23 distances at $\approx$80 ns
(Figure 5.10b top, blue), but there is no corresponding increase in β-structures (Figure 5.10b bottom, green). In the same run, the β-structures do increase to $\approx$20 %, but
only temporarily, and then disappear. The elevated β-structures seen in the reduced
moiety correspond to larger Y/F distances ($\geq$1.5 nm) and not $\approx$1.1 nm as found in the
oxidized moiety. Therefore, it seems that a short Y37/F15 distance, and not Y37/F23,
may initiate the folding to the observed compact state of IAPP, and both Y/F distances
need to be shorter than 1.1 nm in order to stabilize it.

The other variants of IAPP studied do not show such compact structures reached by
the oxidized hIAPP at 330 K; in fact, none of them show β-structure content, as seen in

---

*β-bridges are simply ladders of length one or a residue in an isolated β-bridge, while all other ladders
are labeled as β-ladders or extended strands that participate in a β-ladder. [77]

bottom panels of Figures 5.9 and 5.10. Further investigation of the data, particularly the secondary structure and H-bonds, should shed light on this particular conformation and how it was obtained.

5.2.2 Secondary Structure

5.2.2.1 Ramachandran Angles

Intrapeptide H-bonds and consequently helical secondary structural conformations generally decrease with increasing temperature, indicating greater disordered conformations as temperature increases.[43] Interestingly enough, as illustrated in Chapter 4, the helical content, defined by Ramachandran angles, was more or less constant beyond 330 K and did not decrease continuously as might be expected by comparison with the corresponding H-bond trends. This scenario might be correlated to what seems to be a quasi-2D percolation transition of the hydration water, whose midpoint is about 320 K.[43,112] In other words, the hydration water network around the peptide is more compact below 320 K and limits the freedom of movement of the peptide, whereas above such temperature, the peptide gains flexibility and explores further conformations more easily, thus reducing the helical content of the peptide.[106] The contribution of isolated β-strands and the poly(L-proline) II helices increase slightly with temperature, although their content is almost negligible. This same trend is seen for both moieties of hIAPP, although the trend of the contribution of poly(L-proline) II helices is not as clear as it is for the isolated β-strands. The compact hIAPP structure obtained in this simulation at 330 K presents β-bridges in the C-terminal region. Comparing the Ramachandran plots in Figures 5.11 and 5.12, the decrease of the helical content upon heating corresponds to an increase in the extended conformations, in particular in the isolated β-strand region. The volume under the surfaces delimited by the dashed red lines yield the probabilities of secondary structure listed in Table 5.3. Although this definition may overestimate helical content, especially evident for oxidized hIAPP at 330 K, it allows the calculation of the secondary-structure cooperativity by means of Eq. (4.3). No helix "condensation" was found in these trajectories, most probably owing to the high fluctuations observed in the helices, as can be seen in Figure 5.15, underlining the fact that longer helices are highly transient. In peptides, α-helices can form in 10^{-5} s to 10^{-7} s,[5] so 5×10^{-7} s, i.e., 500 ns, is probably the minimum amount of simulation time required to observe anything significant at 310–330 K, although longer simulation times could reduce the uncertainties, especially at lower temperature; in fact, the first set of 200 ns trajectories, i.e., ALPH described in Section 3.2, turned out to be insufficient.

Table 5.3: *Helical Content assigned by ϕ and ψ angles at 1 bar and at 310 K and 330 K*

	310 K	330 K		310 K	330 K
Ox. hIAPP	39±8	30.2±1.2	Red. hIAPP	47±2	29±14
Ox. rIAPP	33±6	34±4	Red. rIAPP	35.8±1.8	36.0±1.8
Ox. rL23F	41.3±0.8	32.1±1.1	Red. rL23F	36.4±1.7	34±7
Ox. hIAPP$_E$		29.0±0.9	Red. hIAPP$_E$		32±13
Ox. hIAPP$_C$		33.0±0.5	Red. hIAPP$_C$		22.6±1.3

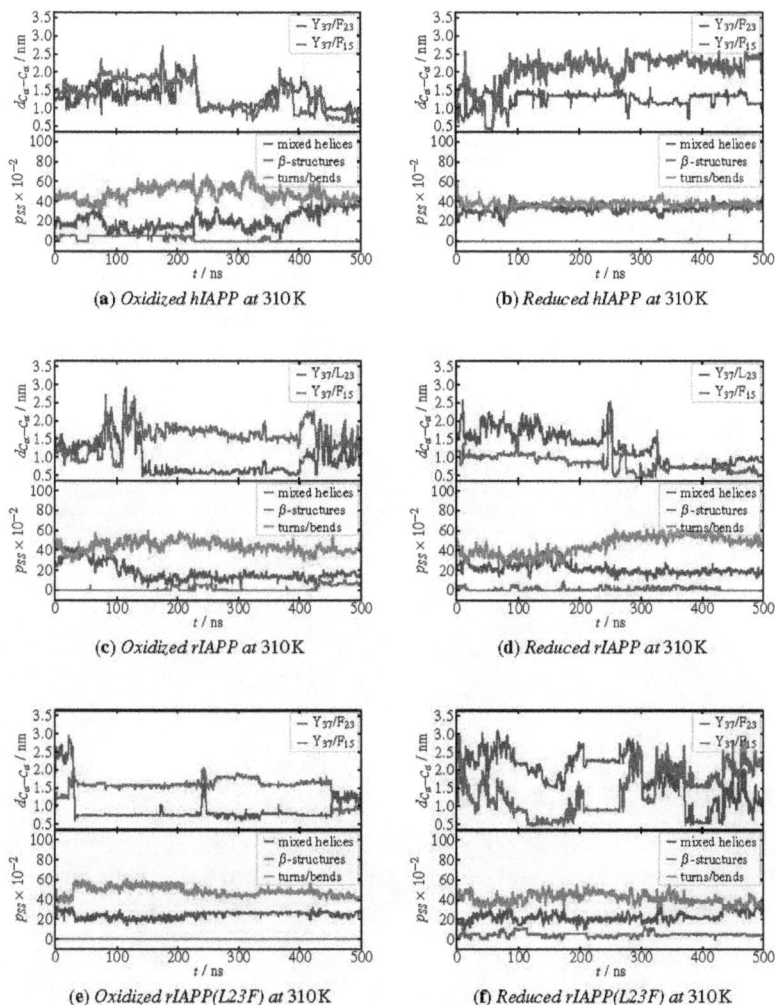

Figure 5.9: *Time dependence of the distance between the Cα, $d_{C\alpha - C\alpha}$, of residues Y37/F23\L23 (in blue, top) and Y37/F15 (in green, top) and secondary structure assigned by DSSPcont,[77,78] where p_{SS} is the probability of secondary structure elements of mixed helices (in blue, bottom), collectively consisting of 3_{10}-helices, α-helices, and π-helices, β-structures (in green, bottom), which comprise β-bridges and β-ladders, and H-bonded turns and bends (in red, bottom). The data temperature is 310 K for (a) oxidized and (b) reduced hIAPP; (c) oxidized and (d) reduced rIAPP; (e) oxidized and (f) reduced rIAPP(L23F).*

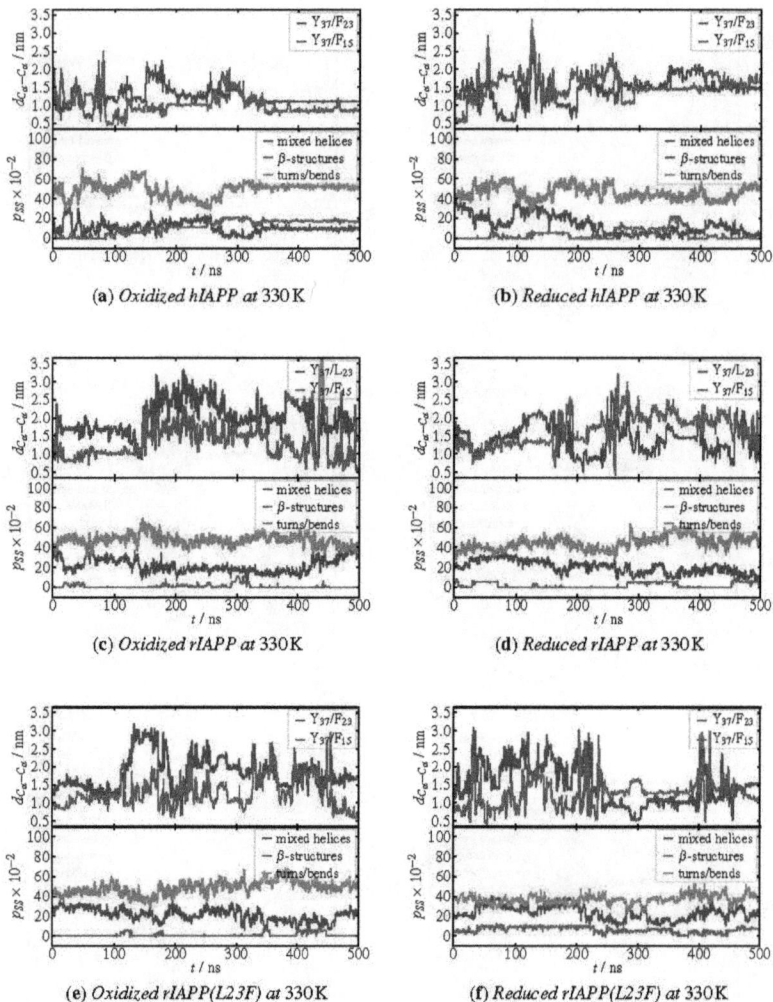

(a) *Oxidized hIAPP at 330 K*

(b) *Reduced hIAPP at 330 K*

(c) *Oxidized rIAPP at 330 K*

(d) *Reduced rIAPP at 330 K*

(e) *Oxidized rIAPP(L23F) at 330 K*

(f) *Reduced rIAPP(L23F) at 330 K*

Figure 5.10: *Time dependence of the distance between the $C\alpha$, $d_{C\alpha-C\alpha}$, of residues Y37/F23\L23 (in blue, top) and Y37/F15 (in green, top) and secondary structure assigned by DSSPcont,[77,78] where p_{SS} is the probability of secondary structure elements of mixed helices (in blue, bottom), collectively consisting of 3_{10}-helices, α-helices, and π-helices, β-structures (in green, bottom), which comprise β-bridges and β-ladders, and H-bonded turns and bends (in red, bottom). The data temperature is 330 K (a) oxidized and (b) reduced hIAPP; (c) oxidized and (d) reduced rIAPP; (e) oxidized and (f) reduced rIAPP(L23F).*

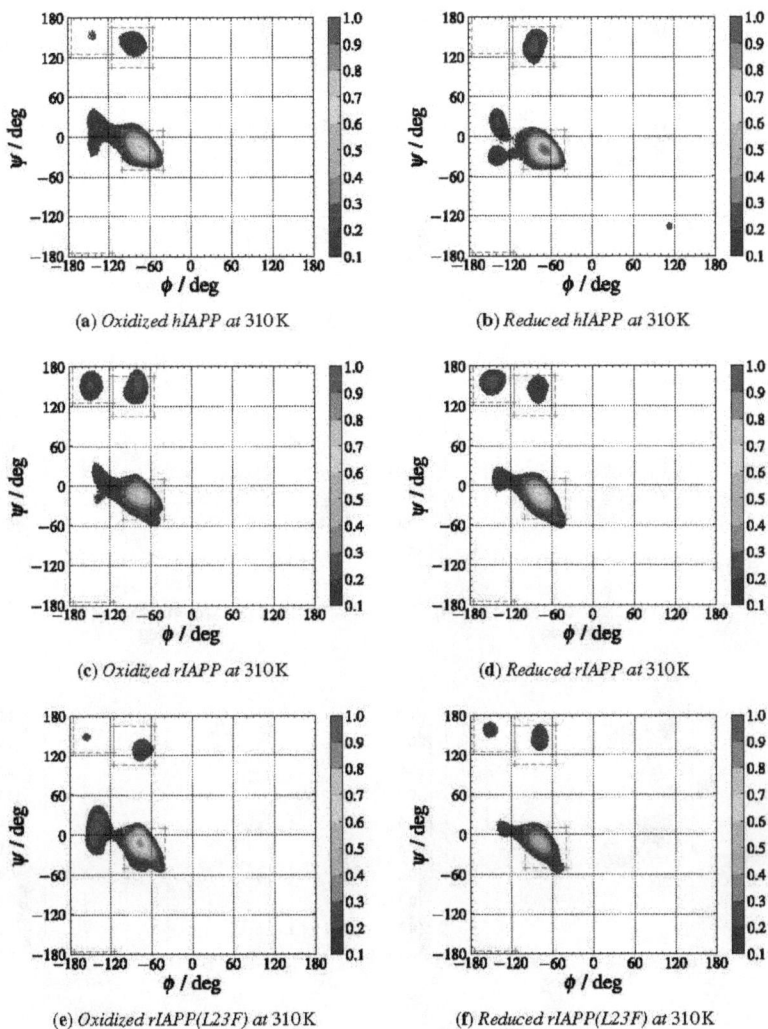

(a) *Oxidized hIAPP at* 310 K

(b) *Reduced hIAPP at* 310 K

(c) *Oxidized rIAPP at* 310 K

(d) *Reduced rIAPP at* 310 K

(e) *Oxidized rIAPP(L23F) at* 310 K

(f) *Reduced rIAPP(L23F) at* 310 K

Figure 5.11: *Ramachandran plots with characteristic secondary structure dihedral angles delimited by dashed red lines, as defined in Section 2.4.1. The occurrence is normalized to Figure 5.11b. The data temperature is 310K for (a) oxidized and (b) reduced hIAPP; (c) oxidized and (d) reduced rIAPP; (e) oxidized and (f) reduced rIAPP(L23F).*

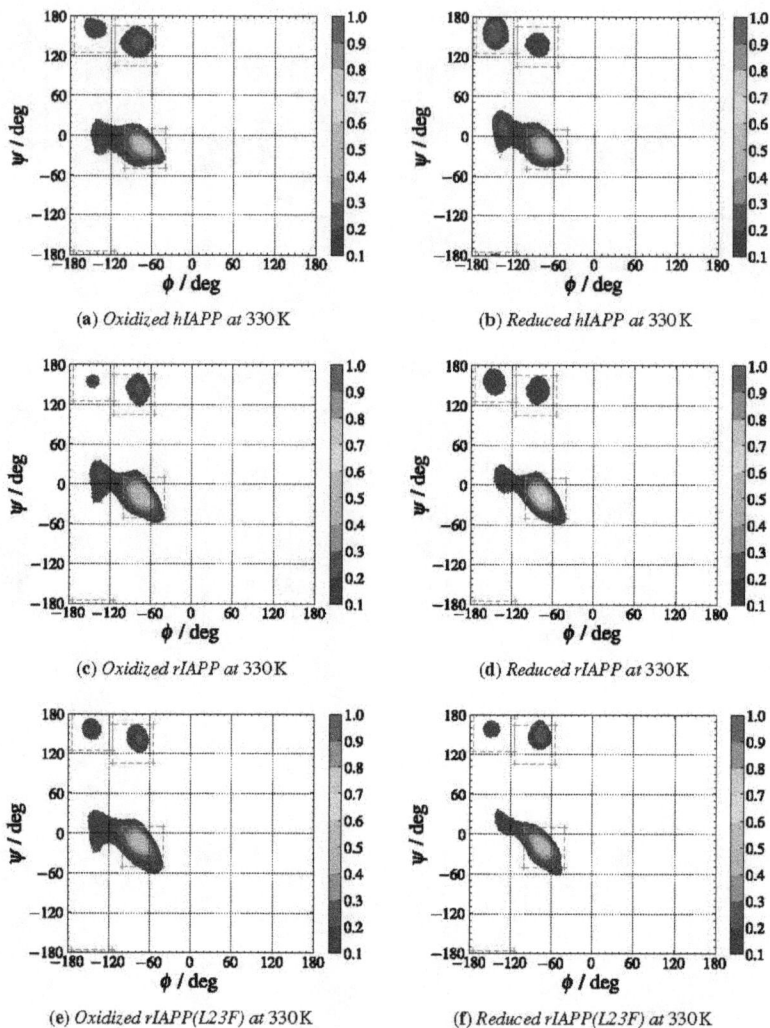

(a) *Oxidized hIAPP at* 330 K

(b) *Reduced hIAPP at* 330 K

(c) *Oxidized rIAPP at* 330 K

(d) *Reduced rIAPP at* 330 K

(e) *Oxidized rIAPP(L23F) at* 330 K

(f) *Reduced rIAPP(L23F) at* 330 K

Figure 5.12: *Ramachandran plots with characteristic secondary structure dihedral angles delimited by dashed red lines, as defined in Section 2.4.1. The occurrence is normalized to Figure 5.11b. The data temperature is 330K for (a) oxidized and (b) reduced hIAPP; (c) oxidized and (d) reduced rIAPP; (e) oxidized and (f) reduced rIAPP(L23F).*

5.2.2.2 DSSPcont

The secondary structure was preferably assigned by DSSPcont,[77,78] and considering the average value of all the helical elements plotted in Figures 5.9 and 5.10, there is a drastic reduction upon heating for both moieties of hIAPP, while rIAPP seems independent of the temperature increase from 310 K to 330 K, as listed in Table 5.4.

Table 5.4: *Helical Content assigned by DSSPcont at* 1 bar *and at* 310 K *and* 330 K

	310 K	330 K		310 K	330 K
Ox. hIAPP	20±10	11±5	Red. hIAPP	33.1±1.1	14±13
Ox. rIAPP	18±12	22±3	Red. rIAPP	21±3	20±5
Ox. rL23F	24.2±1.4	21±3	Red. rL23F	22.5±1.4	24±3
Ox. hIAPP$_E$		11±3	Red. hIAPP$_E$		16±14
Ox. hIAPP$_C$		9.75±0.15	Red. hIAPP$_C$		8±3

The time-dependent plot of collective secondary structure seen in Figures 5.9 and 5.10 in Section 5.2.1 can be seen in detail in Figures 5.13 and 5.14, for 310 K and 330 K respectively. The probability of each residue to be in one of the possible helical conformations, expressed in percentage, can be found in Figure 5.15. hIAPP is clearly disordered in solution at 330 K, as seen in Figure 5.14a, with transient 3_{10}-helices (Figure 5.14a, light blue), β-bridges (Figure 5.14a, violet), and β-ladders (Figure 5.14a, dark blue) throughout the first ≈350 ns of the run. These particular β-ladders are addressed in more detail in Section 5.2.3. At ≈250 ns, the first step towards the formation of β-bridges throughout the C-terminal region occurs through a rearrangement of β-ladders (Figure 5.14a, dark blue) and temporary dissolution of 3_{10}-helices (Figure 5.14a, light blue), occuring approximately 50 ns before the Y37/F23 distance has reached its minimum (blue line in the top panel of Figure 5.10a, at ≈325 ns). At ≈350 ns, the peptide has formed all the aforementioned β-bridges (Figure 5.14a, violet) and does not change for the rest of the simulation run, presenting helical content localized between residues A5 and T10, with the residues contributing from 32 % to 82 % of occurrence; Figure 5.15a depicts data relative to the entire 500 ns, so the average content shown is less than what is measured for the last 150 ns. Lack of helicity between L27 and S29 seems relevant for the flexibility that enables the peptide to fold; in fact, while the peptide is still exploring conformations during the first ≈100 ns, as can be seen by the high fluctuations of R_g (Figure 5.1c, top panel), there is a transient 3_{10}-helix/H-bonded turn in region 27 to 29, not present in the folded conformation. The three β-bridges are all antiparallel and located as follows: A) R11/V17 and A13/L16; B) L16/S28; C) N22/S34. Since the definitions used to assign β-bridges in DSSPcont do not correspond to those used to define H-bonds as described in Section 2.4.2, the results differ slightly from those discussed in Section 5.2.4, see Figure 5.19a. However, the H-bonds calculated by UCSF Chimera[115] for the snapshot of this folded state correspond to the ones assigned in the H-bond matrix (data not shown).

The cysteine-moiety-containing hIAPP at 310 K (Figure 5.13b) shows a transient α-helical content, localized from residues N3 to N21 and from S28 to T30, as seen in the bottom panel of Figure 5.15a in black, with the former helical segment being disrupted by the introduction of the disulfide bond, although transient helical assignments are found in the same region; in fact, the helical content for this oxidized moiety (Figure 5.13a) is localized between residues A5 and V17, with the residues contributing from 6.8 % to 70 %

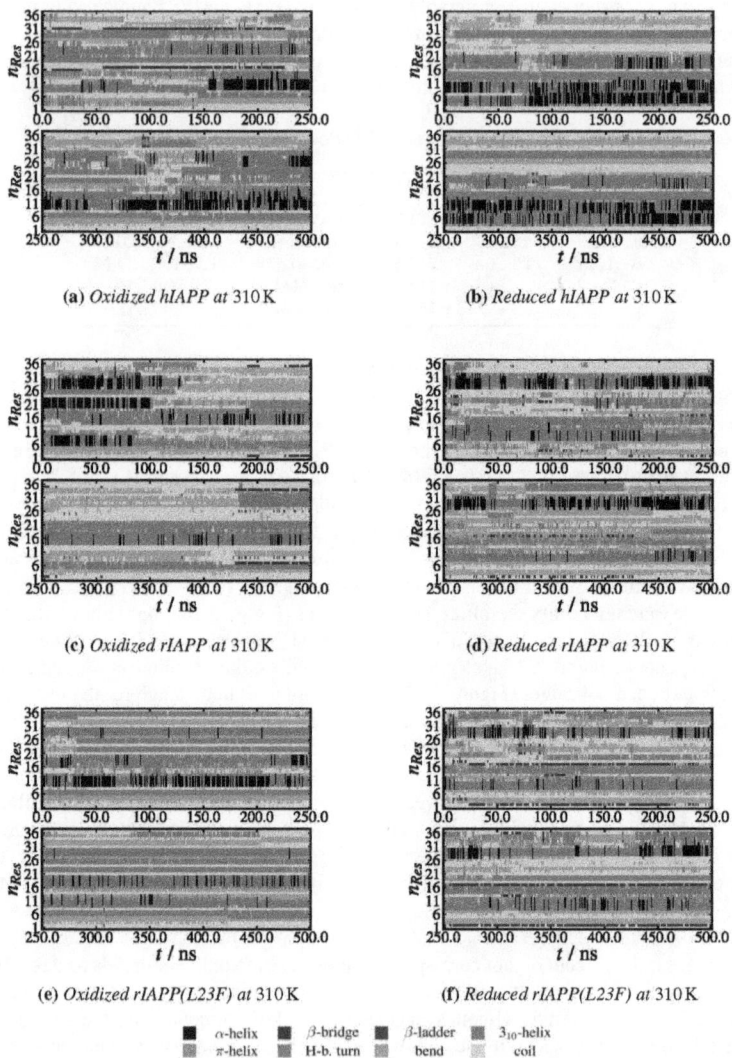

(a) *Oxidized hIAPP at* 310 K

(b) *Reduced hIAPP at* 310 K

(c) *Oxidized rIAPP at* 310 K

(d) *Reduced rIAPP at* 310 K

(e) *Oxidized rIAPP(L23F) at* 310 K

(f) *Reduced rIAPP(L23F) at* 310 K

Figure 5.13: *Time dependence of secondary structure assigned by DSSPcont,[77,78] where 3_{10}-helices are in light blue, α-helices in black, β-bridges in violet, and β-ladders in dark blue; H-bonded turns are in dark green and bends are in light green. The data temperature is 310 K for (a) oxidized and (b) reduced hIAPP; (c) oxidized and (d) reduced rIAPP; (e) oxidized and (f) reduced rIAPP(L23F). The data are plotted every* 1.0 ns.

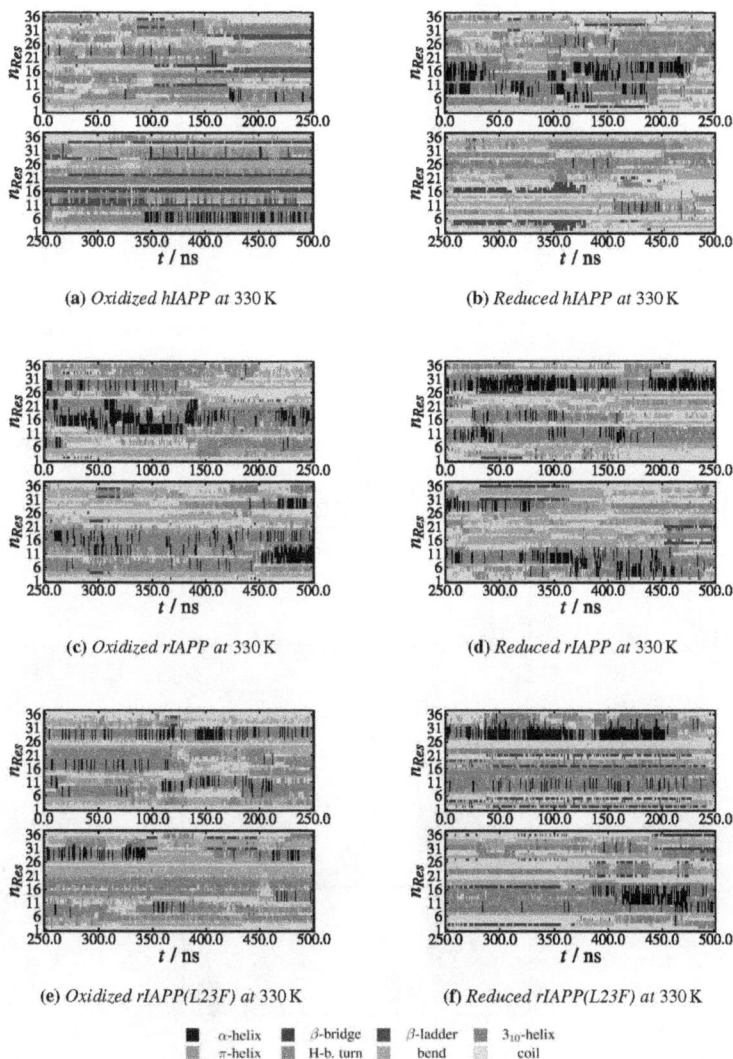

(a) *Oxidized hIAPP at* 330 K

(b) *Reduced hIAPP at* 330 K

(c) *Oxidized rIAPP at* 330 K

(d) *Reduced rIAPP at* 330 K

(e) *Oxidized rIAPP(L23F) at* 330 K

(f) *Reduced rIAPP(L23F) at* 330 K

| α-helix | β-bridge | β-ladder | 3₁₀-helix |
| π-helix | H-b. turn | bend | coil |

Figure 5.14: *Time dependence of secondary structure assigned by DSSPcont,[77,78] where 3₁₀-helices are in light blue, α-helices in black, β-bridges in violet, and β-ladders in dark blue; H-bonded turns are in dark green and bends are in light green. The data temperature is 330 K for (a) oxidized and (b) reduced hIAPP; (c) oxidized and (d) reduced rIAPP; (e) oxidized and (f) reduced rIAPP(L23F). The data are plotted every 1.0 ns.*

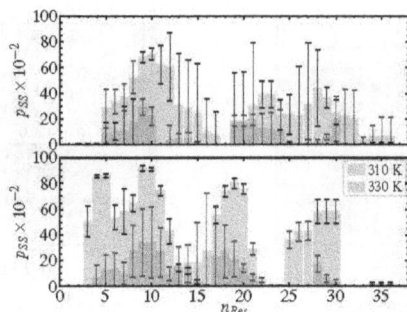

(a) hIAPP at 310 K and 330 K

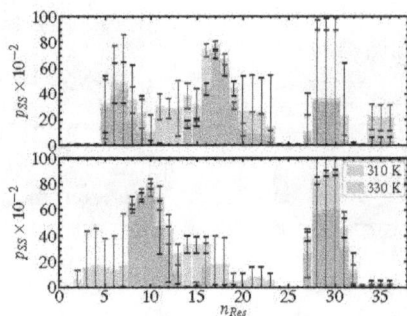

(b) rIAPP at 310 K and 330 K

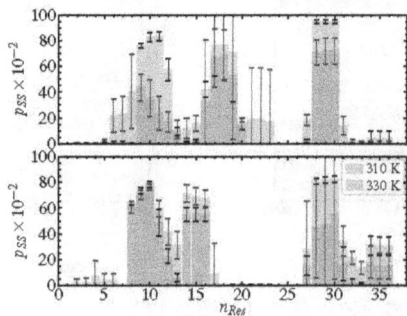

(c) rIAPP(L23F) at 310 K and 330 K

Figure 5.15: *Mean helical content for (a) hIAPP, (b) rIAPP, and (c) rIAPP(L23F) at 310K (in black) and 330K (in blue), as assigned by DSSPcont,[77,78] where any occurrence of 3_{10}, α, or π helices seen in Figures 5.13 and 5.14 contribute to this distribution. The upper panels show the oxidized moiety, while the reduced moiety is found in the lower panels.*

of occurrence (Figure 5.15a, top panel in black). Another transient helix is located between residues S19 and S30, with contribution of the residues between 16 % and 44 %, as seen in Figure 5.15a. Moreover, the temperature dependence of the helical content listed in Table 5.4 can clearly be visualized in Figure 5.15a, where the height of the bars shows a decrease in helicity with temperature. The reduced moiety of hIAPP loses the marked helical content seen at 310 K, with the helical portion of the C-terminus shifting down a few residues, starting from residue twenty-five. In particular, the helical content of the cysteine-moiety-containing peptide disappears almost completely after $\approx$230 ns of the simulation for hIAPP, as seen in Figure 5.14b.

Rodent IAPP shows similar trends to the corresponding human homologue, with a striking difference: rIAPP presents a more persistent helix localized near the proline residues, at L27 and N31, even at 330 K (Figure 5.14c), the temperature at which a similar helix in hIAPP disappears. The H-bonding ability of residues I26 and L27, together with the flexibility in region 25–28, have been shown to define the amyloid-forming potential of the hIAPP(20–29) fragment.[116,117] Therefore, a stable helix in this region, which limits the flexibility and the ability to form interpeptide H-bonds, should also limit the ability to aggregate; in fact, the helical content of the cysteine-moiety-containing peptide disappears $\approx$350 ns for rIAPP (Figure 5.14d), while the cystine-moiety-containing rIAPP (Figure 5.14c) peptide maintains its helical content that starts from around L27 and two regions from A8 and V17. The helical content dissolves in the L27 region at $\approx$125 ns, but reforms at $\approx$375 ns, initially as an H-bonded turn, then as transient $3_{10}/\alpha$-helices. As previously mentioned, a low value of r_{eted} was also observed in oxidized rIAPP at 310 K (blue line in the top panel of Figure 5.2a, between $\approx$120 ns to $\approx$400 ns), approximately in the same time frame in which the helix between L27 and N31 is not present. The regions in which the helicity is present in the mutated rIAPP moieties are quite similar to those observed in wild-type, with the exception of the reduced moiety, which displays a shorter helix between A8 and V17 and is, at the same time, less transient than the wild-type counterpart (Figure 5.15c).

Both the human and rodent homologues share the characteristic of displaying a greater contribution to the helical content of the peptide given by residues 5 through 17. In fact, Nanga et al. have found that the RMSD of the backbone atoms of such interval in rIAPP is (0.22 ± 0.07) Å, while the interval between 5 and 23 yields a backbone atom RMSD of (0.51 ± 0.19) Å.[118] Root-mean-square calculations on these trajectories show a corresponding trend of smaller RMSF for residues 5–17 and larger RMSF for residues 5–23, both of which increase upon heating. The largest RMSF is found in the interval 30–37, which also increases upon heating. The exception to this trend is the oxidized rIAPP at 310 K, but that seems to be due to the fact that P25 is in a *trans* conformation, which could stabilize the helicity in the otherwise more disordered C-terminus. Nanga et al. have seen that P25 and P28, but not P29, undergo *cis-trans* isomerization,[118] but given the slowness of the *cis-trans* isomerization,[5] the transition from the initial *trans* conformation is highly unlikely during MD simulations. In the cysteine moiety rIAPP simulation, P25 underwent a *trans-cis* conformational change. At both temperatures, in the cystine moiety, there were moments in which P25 presented torsion tension for another *trans-cis* isomerization, albeit insufficient for the transition, while P28 and P29 did not present any torsion tension necessary for a *trans-cis* isomerization.

5.2.3 Snapshots of IAPP

Further investigation on this compact conformation and the pathway leading to it shows a short antiparallel β-ladder between residues V17/S28 and L16/S29, with the residues in between forming a loop (as can be seen by the dark blue lines between $\approx$175 ns and $\approx$275 ns in Figure 5.14a and by the H-bonds spawning from S28 and S29 in cornflower blue in Figure 5.16b). This can be compared to the aggregation-prone β-hairpin conformation found by Dupuis et al.;[36] in fact, the snapshot conformation seen in Figure 5.16b collapses quickly into a very compact conformation, but could very well interact with neighboring peptides as a result of the exposure of N^{22}FGAIL^{27} (shown in Figure 5.16b in violet and black ribbon), which was found, along with F^{23}GAIL^{27}, to be the shortest sequence to form β-sheet-containing fibrils.[108] The snapshot in Figure 5.16b was taken at $\approx$220 ns in Figure 5.14a, while the other hIAPP snapshot was taken at $\approx$66 ns, which corresponds to a very expanded conformation with large R_g, as seen in Figure 5.1c.

Not only helical conformations are assigned through DSSPcont, but also H-bonded turns or bends, and in such cases, these proline residues can induce kinks or turns, inhibiting the alignment of elements in the C-terminus necessary for β-structure formation. An example of a bend and of an H-bonded turn may be seen in the snapshots of rIAPP, in Figures 5.16c and 5.16d respectively. These snapshots show the secondary structure elements seen in Figure 5.14c at $\approx$170 ns and $\approx$440 ns. The snapshot of rIAPP (Figure 5.16c) is taken at a time where no defined helical structure around P28 is present; in fact, DSSPcont[77,78] identified it as a bend, where the curvature is at least 70°. The other snapshot (Figure 5.16d) exemplifies how the residues P28/T30 are not quite identifiable as a 3_{10}-helix, but a hydrogen bonded turn, and even such a conformation seems to be sufficient to inhibit the alignment of the residues necessary for the formation of the β-bridges. A snapshot of this proline N-capped α-helix can be seen in the 23–37 fragment of rIAPP on the left of Figure 5.17b.

The main element of secondary structure that can be seen in both moieties is the helix in proximity of C7 (Figure 5.16b and 5.16c).*

An exemplification of what was discussed in Figure 5.10 is the distance between the $C\alpha$ atoms of Y37 (in red) and F23/L23 (in cyan) and residues F15 (in green); in particular, in Figure 5.16b, Y37/F15 is 1.06 nm, and Y37/F23 is 1.22 nm. Besides the depicted peptide in Figure 5.16c that presents a Y37/F15 distance of 1.6 nm, while Y37/L23 is 2.6 nm, all the other peptides show $d_{C\alpha-C\alpha}$ less than 2.0 nm; hence leading to a compact conformation compared to random walk distances of at least 3.0 nm or 4.0 nm in either case.[30]

Additional snapshots of a few characteristics of IAPP can be seen in Figure 5.17. The N-terminal region of hIAPP has been shown to interact with negatively charged membranes undergoing an α-helix to β-sheet conformational change,[8] along with other studies that hypothesize biological activity with the helicity of the N-terminal region;[21] therefore, a snapshot of this α-helical region from residues 8 to 22 was prepared to show the hydrophobic/hydrophilic character of IAPP. In Figure 5.17a, one can see that one side of the peptide is clearly hydrophobic (on the left, in orange), while the other is clearly hydrophilic (on the right, in blue). This region is exactly like rIAPP, with the exception of the eighteenth residue, and both interact with the membranes,[29] so the distribution of the peptides would suggest there is a putative binding site with hydrophobic or hydrophilic regions. This is purely speculative, but it could be interesting to investigate

---

*Confirmed by a private communication with Prof. Dr. David Eisenberg at "Amyloid 2009" in Halle.

(a) *Oxidized hIAPP at ≈66ns*

(b) *Oxidized hIAPP at ≈220ns*

(c) *Oxidized rIAPP at ≈170ns*

(d) *Oxidized rIAPP at ≈440ns*

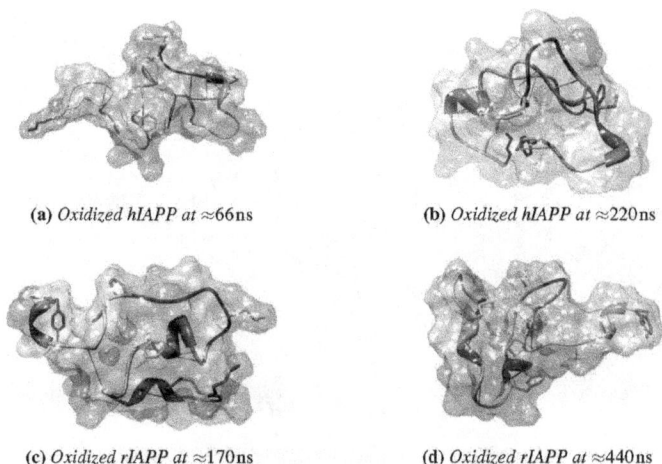

Figure 5.16: *Snapshots of oxidized (a)–(b) hIAPP and (c)–(d) rIAPP at 330K, where K1 is blue, C2 and C7 are in yellow, Y37-NH$_2$ are red, F15 is green, F23/L23 is cyan, and S28/P28 and S29/P29 are in cornflower blue. The violet ribbon corresponds to the $N^{22}(F/L)GAIL^{27}$ sequence, with G24 in black. The POV-Ray rendered[119] images are made with UCSF Chimera;[115] the orange helices and the overall ribbon representation of the secondary structure are assigned by ksdssp,[77] and the surface was calculated with the MSMS package.[120] The surface shows the amino acid hydrophobicity in the Kyte-Doolittle scale[121] with colors ranging from dodger blue for the most hydrophilic to orange red for the most hydrophobic, and white at 0.0.*

this α-helical region of IAPP and membranes, or IAPP with possible binding proteins, via MD simulations. The snapshot of oxidized rIAPP at ≈26ns of Figure 5.13c is characterized by an interaction between the helix observed in the N-terminal region (on the left) and the C-terminal region (on the right), as can be seen in Figure 5.17b. Bringing these two fragments of the peptide together, one can see the overlap of the corresponding hydrophobic-hydrophobic and hydrophilic-hydrophilic regions defined by the exposure of side chains in the helices. This conformation could be one of the possible stable conformations of rIAPP in solution that is less compact than the β-ladder-rich hIAPP monomer because of the higher helical content.

5.2.4   H-bond Patterns and Secondary Structure of Oxidized hIAPP at 330 K

The distribution of average backbone-backbone H-bonds between residues (Figures 5.18 and 5.19) may also be used to localize secondary structure elements, by the position of H-bonds on the matrix as described in Section 2.4.2 on page 30.

In Figure 5.19a, a detailed residue-residue H-bond matrix of the last 150 ns of the run of oxidized hIAPP at 330 K is depicted. A short helix can be seen in the N-terminus, between R11 and T4, from L16 to A13, and from N21/N22 to H18. An important point, present only in the folded peptide, is T36/C2, which also corresponds to a short r_{eted}.

(a) *Ox. hIAPP α-helix residues 8–22*

(b) *Ox. hIAPP α-helix residues 8–22(r) and 23–37(l)*

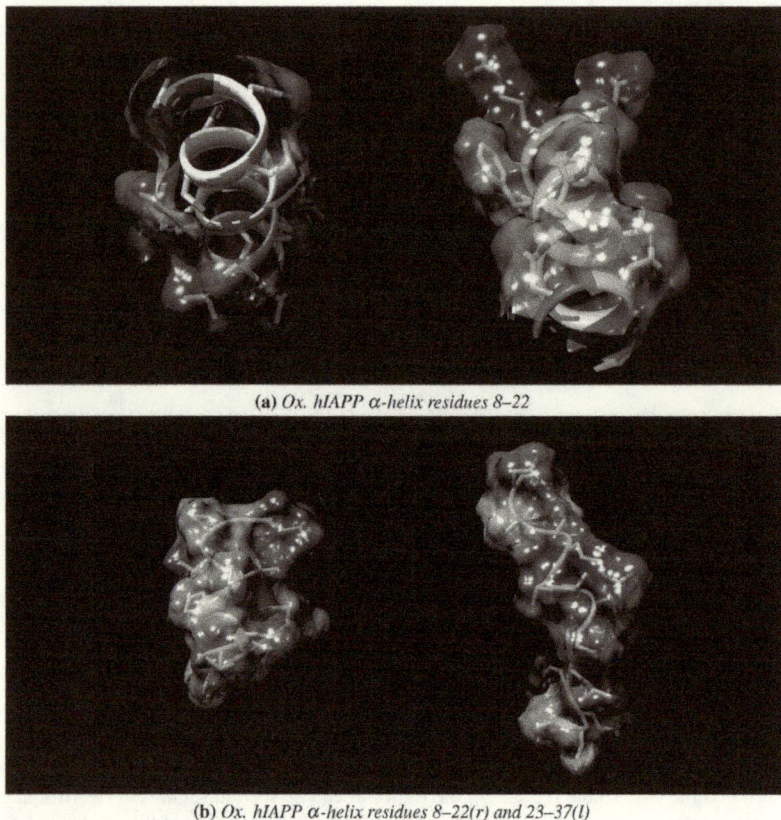

Figure 5.17: *(a) The hIAPP region 8 to 22, with A8 (in red ribbon) pointing towards the reader and N22 (in blue ribbon) pointing away from the reader, shows hydrophobic residues on the bottom of the peptide (left) and hydrophilic residues on top (right). The surface shows the amino acid hydrophobicity in the Kyte-Doolittle scale[121] with colors ranging from dodger blue for the most hydrophilic to orange red for the most hydrophobic, and white at 0.0. The white ribbons show the initial α-helical conformation that was simulated at 350 K, as described in Section 2.2. (b) Oxidized rIAPP at ≈26 ns at 310 K showing the helical fragment 23 to 37 (left) and 8 to 22 (right) with A8 pointing downwards. The two parts are taken from the same conformation, which has been split in order to show the hydrophobic and hydrophilic interactions between the N-terminal and C-terminal halves. Please note the proline N-capped α-helix in the figure on the left. The POV-Ray rendered[119] images are made with UCSF Chimera;[115] the orange helices and the overall ribbon representation of the secondary structure are assigned by ksdssp,[77] and the surface was calculated with the MSMS package.[120]*

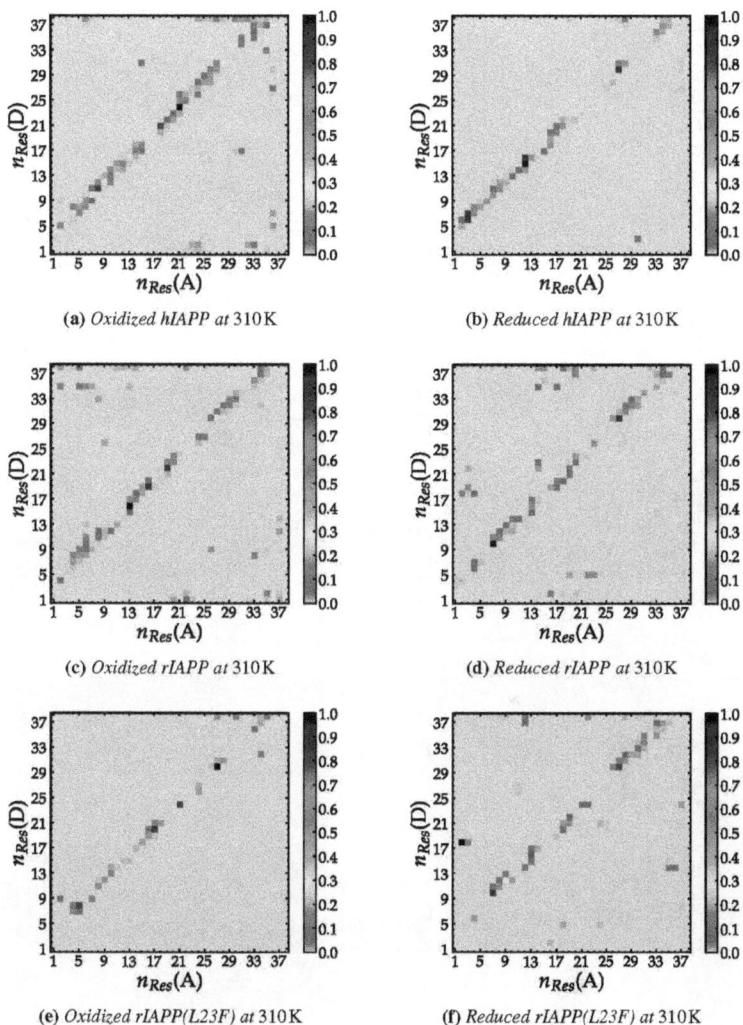

(a) *Oxidized hIAPP at* 310 K

(b) *Reduced hIAPP at* 310 K

(c) *Oxidized rIAPP at* 310 K

(d) *Reduced rIAPP at* 310 K

(e) *Oxidized rIAPP(L23F) at* 310 K

(f) *Reduced rIAPP(L23F) at* 310 K

Figure 5.18: *Normalized distribution of the average backbone-backbone H-bond plots between residues, where the residue number, n_{Res}, on the abscissa is the acceptor (A) of the pair and the ordinate is the donor (D). Residue 38 is the amide cap of Y37. The occurrence is depicted by a ROYGBIV spectral color code and increases as red shifts to violet, as the colors of the white light spectrum; the minimum and maximum are gray and black, respectively. The data temperature is 310 K for (a) oxidized and (b) reduced hIAPP; (c) oxidized and (d) reduced rIAPP; (e) oxidized and (f) reduced rIAPP(L23F). All the data presented were obtained from the entire 500 ns simulation runs.*

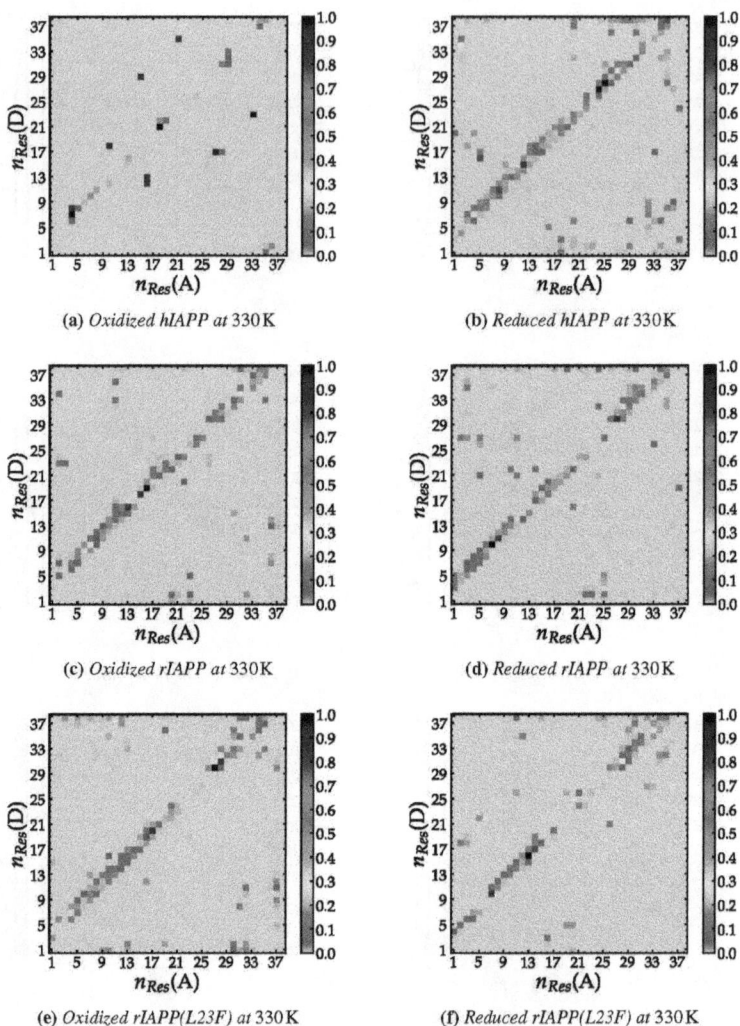

(a) *Oxidized hIAPP at* 330 K

(b) *Reduced hIAPP at* 330 K

(c) *Oxidized rIAPP at* 330 K

(d) *Reduced rIAPP at* 330 K

(e) *Oxidized rIAPP(L23F) at* 330 K

(f) *Reduced rIAPP(L23F) at* 330 K

Figure 5.19: *Normalized distribution of the average backbone-backbone H-bond plots between residues, where the residue number, n_{Res}, on the abscissa is the acceptor (A) of the pair and the ordinate is the donor (D). Residue 38 is the amide cap of Y37. The occurrence is depicted by a ROYGBIV spectral color code and increases as red shifts to violet, as the colors of the white light spectrum; the minimum and maximum are gray and black, respectively. The data temperature is 330 K (a) oxidized and (b) reduced hIAPP; (c) oxidized and (d) reduced rIAPP; (e) oxidized and (f) reduced rIAPP(L23F). Only the last 150 ns are represented in (a); the entire 500 ns, in all the other runs.*

Other relevant points on the H-bond matrix, with occurrences above 80 % (Figure 5.19a, from dark blue to black), are V17/L27, H18/Q10, F23/G33, S29/F15, and N35/N21. The N21/H18 helix is present throughout the 330 K run, as are the H18/Q10 and the aforementioned β-bridges except S29/F15, albeit to a lesser extent (data for 500 ns not shown). What is interesting to note is that H18 also participates in a β-bridge with Q10, but there is no corresponding equivalent H-bond in the rIAPP data involving R18.

At 310 K, the cystine moiety presents helices from H18 to T4 and from N31 to H18, traces of diagonal elements related to β-bridges, and a relatively high mobility of the last residues, which form many H-bonds, albeit with low occurrence (Figure 5.18a). The cysteine moiety, on the other hand, shows a more prominent helical content between N22 and C2, with the C-terminal region not as mobile as seen in the oxidized counterpart at 310 K (Figure 5.18b). Another persistent helical element involves T30/L27, which shifts towards S28/G24 upon heating (Figure 5.19b). At 330 K, the reduced moiety of hIAPP presents many transient helices, a flexible C-terminus, with fluctuating H-bonds along the perpendicular diagonal, none of which correspond to the critical points seen for the oxidized moiety.

Figure 5.19c illustrates the H-bond matrix of oxidized rIAPP at 330 K, where a strong, albeit fluctuating, helical component is found between S19 and A5, with the highest occurrence in S19/L16* and R18/F15, without any of the β-bridge elements seen for the human homologue. The C-terminal region fluctuates, but limited to transient helices, with far less points off the main diagonal as seen for the corresponding hIAPP moiety. In fact, none of the previously mentioned critical points are seen in rIAPP, especially the two H-bonds formed by H18. The H-bond matrix of rIAPP(L23F) is very similar to rIAPP, with an extremely high occurrence of the helix involving N31/L27, even at 330 K (Figures 5.18e and 5.19e). The mutant differs in the presence of H-bonds on the diagonal G24/N21 at 310 K (Figure 5.18e) and F23/S20 at 330 K (Figure 5.19e).

5.2.5 System Perturbation

5.2.5.1 Thermal Induced "Unfolding"

Even after extending the simulation run by 100 ns, of which only the first 10 ns are shown as reference and labeled as COMP, the compact conformation does not "unfold." Only upon heating does the system show increase of R_g, as seen in Figure 5.20a, and r_{eted}, as seen in Figure 5.20c. Although the peptide seems to open up to less compact conformations at 390 K, it still is pretty compact at 410 K. This is most definitely due to the poor sampling; even though the simulations are performed at high temperature, they are still only 10 ns long. Obviously, at 450 K the peptide samples many extended conformations, reaching values of r_{eted} that are greater than 4 nm. As a first approximation, the compact peptide "unfolds" at 390 K, but further investigation at lower temperatures could lower this limit.

The perturbation on the helical content of the peptide is seen in Figure 5.21a. The peptide is clearly unstructured at 450 K. At lower temperatures, i.e., between 350 K and 410 K, the helical content of the peptide seems to increase between residues 5 to 10 and corresponds to an increase in r_{eted}. Similar perturbations are seen when introducing single point mutations into the "folded" structure, as illustrated in Section 5.2.5.2.

---

*Errata corrige: in Ref. 44, it should read L16 instead of L18.

5.2.5.2   In silico Point Mutations on Oxidized hIAPP at $330\,\text{K}$

In silico mutations performed on the collapsed oxidized hIAPP conformation, S28P, S29P, and G24P, disrupt its compactness. In fact, as can be seen in Figure 5.16b, S28 and S29 (side chain shown in cornflower blue), which participate in the formation of a β-sheet, and G24 (depicted by a black ribbon), whose flexibility helps in the formation of the loop between H18 and L27, would be strongly perturbed by the substitution of proline. The proline residues in the serine-to-proline single point mutations are present in wild-type rIAPP, while the glycine-to-proline is not. Short 10 ns simulations performed on these in silico mutations disrupt the compact "folded" conformation obtained at 330 K, as can be seen in Figures 5.20b and 5.20d. Both G24P and S29P show a broadening of the distribution to larger values of R_g, while S28P shows a broad shoulder, with a tail that is given by the "unfolding" of the collapsed state. The perturbation is seen more clearly for r_{eted} distributions, with the lowering of the distribution peak and the formation of a shoulder in G24P and S29P and a tail in S28P.

The small increase in R_g in the G24P mutation trajectory is caused by the disruption of the secondary structure of the C-terminus, as can be seen in the increase in helicity in Figure 5.21b, which induces fluctuations in the Y37/F23 distances (data not shown). S29P perturbs the Y37/F15 distances, from 0.87 nm to 1.02 nm, as well as the increase in R_g. What is most surprising is the perturbation of the S28P mutation. It does not perturb the β-structure around it, as one might expect, rather it disrupts the end-to-end distance, most probably by inducing torsion that detaches the C-terminus from the disulfide region; in fact, the H-bond between T36/C2, shown in Figure 5.19a, decreases from $\approx 60\%$ to $\approx 10\%$ (data not shown). T36 then forms an H-bond with the amide terminus of Y37, with an occurrence of $\approx 35\%$. Both S28P and S29P, as seen in Figure 5.21b, increase the mixed helical content between residues 29 to 31 from 13% to 55% and from 13% to 41%, respectively. Hence, these perturbations induced by serine-to-proline mutation, which destabilize the compact conformation in less than 10 ns of simulation, may very well be sufficient to induce rigidity in rIAPP, thus inhibiting the polypeptide to reach a short end-to-end distance, which seems necessary to reach compact states[35] or states that present aggregation-prone β-sheets.[36]

5.3   Discussion and Conclusions

Single peptide runs obviously cannot reveal mechanistic information about the aggregation process, but they can unravel conformations and regions of the peptide responsible for initiation and inhibition of the aggregation reaction of IAPP, as seen for the helix around P28, the absence of aromatic residues in key positions such as F23 in native rIAPP, and the presence or absence of the disulfide bond in hIAPP.

5.3.1   Compact, but not Entirely Disordered, Polypeptide

Either moiety of hIAPP forms a more or less compact structure, displaying a small radius of gyration in either case, albeit different in secondary structure content, with the natural cystine moiety yielding a more "flexible" peptide, particularly above the percolation transition at 320 K. In fact, the oxidized moiety of hIAPP at 330 K collapses into a com-

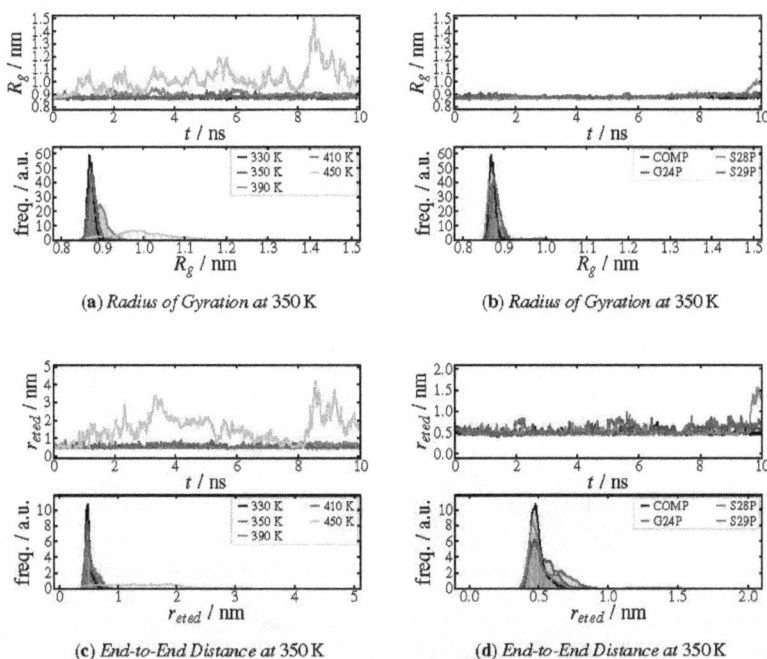

(a) *Radius of Gyration at 350 K*

(b) *Radius of Gyration at 350 K*

(c) *End-to-End Distance at 350 K*

(d) *End-to-End Distance at 350 K*

Figure 5.20: *The time-dependent data of R_g for (a) temperature perturbation of the compact conformation COMP (from 330 K to 450 K) and (b) in silico mutations G24P, S28P, and S29P at 330 K, with a frequency distribution in the lower panels. The same display of data can be seen for r_{eted}, for (c) temperature perturbation of the compact conformation and (d) in silico mutations.*

pact state about 10 % smaller than both rIAPP moieties and the reduced hIAPP moiety. The presence or absence of helical conformations, even if transient, may influence the kinetics of the aggregation. Experimental results from Koo and Miranker indicate that the aggregation is faster when the disulfide bond is present.[34] Moreover, Padrick and Miranker have shown that the aggregated peptide has an ordered C-terminus, which is less ordered when in solution.[27] Padrick and Miranker have also shown that the tyrosine and phenylalanine residues are within 1.5 nm in either homologue of IAPP.[27] The mean distance between $C\alpha$–$C\alpha$ of Y37/F23 and Y37/F15 that were measured, fluctuates greatly between 0.5 nm and 2.5 nm, in which many instances can be efficiently measured with FRET, thus confirming that with such tyrosine-phenylalanine measurable distances the peptide is not as disordered as an unfolded peptide would be.

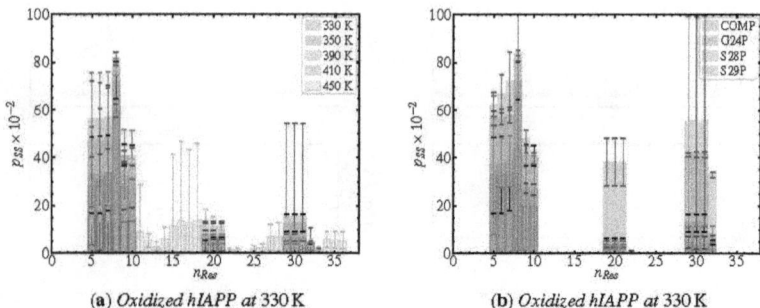

(a) *Oxidized hIAPP at* 330 K   (b) *Oxidized hIAPP at* 330 K

Figure 5.21: *Mean helical content for (a) temperature system perturbation and (b) in silico mutations, as assigned by DSSPcont,[77,78] where any occurrence of* 3_{10}, α, *or* π *helices contribute to this distribution. A reference MD simulation continuation run on the compact conformation obtained at* 330 K *is labeled COMP (in black).*

5.3.2   Effect of P28 on the C-Terminal Region

The MD simulations of monomeric human and rodent IAPP show an important aspect of the C-terminal region: It seems to be capable of forming helices starting from residue 28 when proline is present, i.e., rIAPP and the in silico mutants hIAPP(S28P) and hIAPP(S29P). In fact, both rIAPP and rIAPP(L23F) form helices in this region, a region that is highly unordered in oxidized hIAPP at 330 K. This is possibly due to the nature of residue 28, one of the six different residues in sequence between human and rodent IAPP. Westermark et al. have shown that the S28-for-P28 substitution inhibits the aggregation greatly.[13] Kayed et al. have shown that the polypeptides in solution are prevalently random coil in nature;[10] this was confirmed by the low content of β-structures and a transient helical content seen in the computer simulations.

The helical region around residue P28 is probably not the only reason why rIAPP does not aggregate, but it definitely is relevant. In fact, P28 and the ordered structure around it could limit the interactions between the C-terminal halves of neighboring peptides, especially the H-bonding I26 and L27 residues, and the aromatic residues that seem fundamental for the aggregation process. The examination of the H-bond matrix shows that S28 is enclosed by its two neighboring residues that form pairs, V17/L27 and S29/F23, which seem to stabilize the compact conformation characterized by three antiparallel β-bridges, yielding thus an average R_g value that is at least 9 % lower than those calculated for the other variants of IAPP. In fact, Soong et al. have also shown that hIAPP is significantly more compact than rIAPP, suggesting a higher degree of folding.[113] The MD simulations show a stable helix in this region, other than bends or H-bonded turns, limiting its flexibility and the ability to form interpeptide H-bonds and, thus, should also limit its ability to aggregate. In fact, the oxidized rIAPP run at 310 K presented a short end-to-end distance, which seems necessary for the peptide "folding," only when the helix around S28 was not present.

5.3.3   Effect of Aromatic Residues

The difference in amyloidogenicity between human and rodent amylin could be caused by the presence of aromatic residues along the C-terminal half. In fact, rIAPP(L23F) also aggregates, albeit at low yields.[14] However, P28 may limit the interaction of the C-terminal halves that seems relevant for interaction and aggregation. In fact, residues I26 and L27 and/or the flexibility in region 25–28 may define the amyloid-forming potential of the hIAPP(20–29) fragment, as shown by Moriarty and Raleigh[116] and Azriel and Gazit.[117]

The in silico mutation of oxidized rIAPP reveals a helical region around P28 throughout the entire 500 ns MD simulations at 310 K and 330 K. Little or no secondary structure elements are located in proximity of residue F23, rendering residue 23 in the oxidized L23F mutant mobile, allowing it to switch between being exposed to solvent and being embedded within the peptide. As a result, a facilitated interaction with neighboring aromatic residues could initiate aggregate formation. Moreover, the H-bond pair F23/G33, contributing to stabilize the β-bridges, involves a phenylalanine residue, which could require Y37 to align the participating residues in order to initiate the formation of the β-bridges. In fact, exposure of hydrophobic patches seems to be relevant for the formation of the β-bridges, since the hydrophobic surfaces diminish with the total SASA and two clear states can be seen in the distribution of such values. Moreover, phenylalanine has been found to enhance β-sheet formation.[2] Various other elements should be considered for determining the amyloidogenicity of IAPP. The presence of P28 in rIAPP definitely influences the secondary structure of the C-terminal half. The high mobility of F23 in rIAPP(L23F) could be the driving force of the aggregation of the L23F mutant, while the structure around P28 could be inhibiting, limiting the yield of the aggregate. Such helicity, albeit transient, may limit the mobility of residue F23 and inhibit, therefore, aggregate formation. F23 in IAPP also seems relevant for initiating the "folding" to a compact conformation. In fact, a short Y37/L23 observed in rIAPP at 310 K and a short end-to-end distance, which seems necessary for folding, did not lead to a collapse possibly for the lack of Y37/F23 aromatic-aromatic interaction observed for hIAPP.

5.3.4   Temperature Effect on Oxidized hIAPP

Kayed et al. have also shown that hIAPP, when heated from 298 K to 363 K, undergoes thermal "denaturation." In particular, a highly cooperative conformational transition occurs when the temperature was raised from 308 K to 318 K,[10] which compares well with the temperature chosen for these calculations. The percolation transition for IAPP has been calculated at $\approx$320 K,[43,112] so studying the monomeric system at 310 K and 330 K should shed light on the mechanism of a conformational transition that occurs during the lag phase prior to aggregation of hIAPP. In fact, all the runs at 310 K, with the exception of the hIAPP cystine moiety, show little or no exploration of such possible states. Moreover, at 310 K reduced hIAPP presents also a relatively stable helix starting from S28, which could be one of the causes of a slower aggregation of the cysteine moiety. On the other hand, at 330 K oxidized hIAPP is flexible enough to allow the termini to be within $\approx$0.54 nm and form β-bridges and β-ladders in the C-terminal region, with the contemporary formation of stable, albeit transient, helices in the N-terminus around C7. More recent studies on the temperature dependent aggregation by Vaiana et al. have shown that the monomeric form, at lower temperatures, could be detected for long periods of time, while aggregation appeared immediately at higher temperature, [122] confirming the low reactivity of hIAPP at lower temperatures. Moreover, the reactive conformation of oxidized hIAPP, found to collapse at 330 K compared to the other relatively inert conformations that did not collapse, is confirmed by the fact that Kayed et al. also found two distinct conformers presenting different amyloidogenic properties, [10] later confirmed by Dupuis et al. [36]

5.3.5 Effect of the Disulfide Bond

The presence of the disulfide bond in hIAPP makes the peptide more flexible and able to sample more conformations in the C-terminal half, thus facilitating interactions with neighboring peptides, as can also be seen in transient isolated β-strands localized in the C-terminus (data not shown). Moreover, the disulfide of the cystine seems to stabilize the short end-to-end distance in the oxidized moiety of hIAPP, [35] allowing the formation of aggregation-prone β-sheets. [36] The presence of disulfide bond allowed the observation of a short end-to-end distance in rIAPP, although a subsequent folding of the peptide did not occur. On the other hand, the absence of the disulfide bond also influences the stability of the peptide. In fact, the peptide shows cooperative helicity, especially at lower temperatures, between residues 4 and 22, hence less flexible. In contrast, no stabilizing effect of the disulfide, which was found to form a template that stabilizes the short end-to-end distance, [35] was observed at higher temperatures, where the turn content increases, leaving the peptide completely unstructured.

Chapter 6

Outlook

Studying an unstructured peptide in solution is definitely a daunting task, but nevertheless some characteristics of IAPP were identified. Three seemingly important conditions need to be met in order to observe folding of hIAPP at 330 K: a) short end-to-end distance, b) short Y/F distance, and c) β-structure H-bond formation. However, further investigation of the order in which they take place is needed to fully understand the underlying mechanism that brings IAPP to a state $\approx 10\%$ more compact than rIAPP in solution.

The most obvious point is the presence of the disulfide bond, which makes the peptide more flexible than the reduced moiety and apparently stabilizes short end-to-end distances,[35] necessary for IAPP folding.

Aromatic residues also seem to play an important role in allowing the peptide to reach this compact conformation, although it is still unclear which interaction initiates the folding. The most obvious interaction occurs when Y37 and F23 come closer, allowing the peptide to reduce its radius of gyration brought about by the increase in β-ladders to $\approx 20\%$ in the normally unstructured C-terminal region of hIAPP. However, the first formation of β-structure H-bonds occurs when Y37 and F15 are within 1.2 nm, so further investigation on the importance of this step in the folding process is needed. In fact, a loss of entropy occurs in molecular interactions because the degrees of freedom are reduced when two parts are brought together. Many such interactions can actually stabilize the folded conformation, and the entropy loss for consecutive interactions can be less than the sum of individual interactions.[5] In other words, bringing two parts together initially requires energy for the entropy loss, but the successive steps require less energy because the interacting parts are kept close thanks to the previously formed interactions. Therefore, further investigation is needed in order to clarify whether the formation of the initial β-structures may occur even without this first aromatic-aromatic interaction, given by short Y37/F15 distances.

Another point that would need clarification is how the rIAPP(L23F) in vitro mutant aggregates.[14] Unfortunately, no particular behavior of the monomeric in silico mutant was observed, if not for a stable helix around P28, although a particular conformation showing a short end-to-end distance was seen for the oxidized wild-type rIAPP, when this helix around P28 was absent. Such conformation was characterized by a short Y37/L23 distance, but being leucine aliphatic, it was not possible to observe any aromatic-aromatic interactions. One possibility for further investigation would be to produce an L23F in silico mutation on such conformation observed in rIAPP and calculate a trajectory starting from that perturbed conformation (obviously, after appropriate minimization). Another possibility would be to perform MD simulations on the rIAPP(L23F) mutant by restraining the Y37/F23 distance to 1.2 nm, while allowing the peptide to explore the conformations limiting the degrees of freedom. Since the in vitro mutation formed fibrils, albeit in low yield, it could be helpful to observe the effect proline, in particular P28, has on the secondary structure.

Any possible combination of residue 25 in silico mutations on monomeric IAPP peptides could be interesting, in virtue of its hypervariability in the various homologues compared to the presence of proline in rodent IAPP.

Further investigation could be taking the snapshot conformation seen in Figure 5.16b and perform MD simulations on oligomeric forms in solution, investigating the interactions of the $N^{22}FGAIL^{27}$ sequence between neighboring peptides. Since that particular conformation collapsed onto itself, it would be interesting to see what it would do with similar neighboring conformations.

Finally, it would be interesting to investigate how various conformations of this monomeric peptide obtained in this work, and in particular the helical portion between residues 8 to 22, interact with model lipid bilayers.

Appendix A

Proceedings

John von Neumann Institute for Computing

NIC

Effect of Temperature on the Structural and Hydrational Properties of Human Islet Amyloid Polypeptide in Water

M. N. Andrews, I. Brovchenko, R. Winter

published in

From Computational Biophysics to Systems Biology (CBSB08),
Proceedings of the NIC Workshop 2008,
Ulrich H. E. Hansmann, Jan H. Meinke, Sandipan Mohanty,
Walter Nadler, Olav Zimmermann (Editors),
John von Neumann Institute for Computing, Jülich,
NIC Series, Vol. 40, ISBN 978-3-9810843-6-8, pp. 153-156, 2008.

http://www.fz-juelich.de/nic-series/volume40

Effect of Temperature on the Structural and Hydrational Properties of Human Islet Amyloid Polypeptide in Water

Maximilian N. Andrews, Ivan Brovchenko, and Roland Winter

Department of Chemistry, Biophysical Chemistry, Dortmund University of Technology,
D-44227 Dortmund, Germany
E-mail: {maximilian.andrews, ivan.brovchenko, roland.winter}@tu-dortmund.de

Structural and hydrational properties of the full-length human islet amyloid polypeptide 1-37 (hIAPP) were studied in a temperature range from 250 to 450 K by molecular dynamics computer simulations. At all temperatures studied, hIAPP does not adopt a well-defined conformation. The distribution of residues having the dihedral angles ϕ and ψ within the allowed regions of the Ramachandran plot which define β-sheets and poly(L-proline) II structures along the peptide chain is close to random, whereas a clear trend towards cooperative "condensation" is seen for residues having Ramachandran angles which characterize α-helices. This cooperativity and the number of intrapeptide H-Bonds is suppressed by heating or by introducing the natural intramolecular disulfide bond between residues 2 and 7. Intrinsic volumetric properties of hIAPP were estimated by taking into account the difference in the volumetric properties of hydration and bulk water. The temperature dependence of the density of hydration water indicates that the effective hydrophobicity of the hIAPP surface is close to that of carbon-like surfaces. The thermal expansion coefficient of hIAPP is found to be negative and decreases continuously upon heating from $\sim -3 \cdot 10^{-4}$ to $\sim -2 \cdot 10^{-3}$ K^{-1}. The spanning H-bonded network of hydration water at the hIAPP surface breaks via a percolation transition at about 320 K, which may be related to the drastic speed up of hIAPP aggregation seen experimentally in this temperature region.

1 Introduction

The aggregation of the human islet amyloid polypeptide (hIAPP) is involved in Diabetes Mellitus Type II. Hence, knowledge of the conformational behavior of this peptide is important for understanding the aggregation mechanism of hIAPP and for finding the means to prevent formation of its ordered fibrillar aggregates, which may be the main cause of decease. Experimental studies of the structural properties of hIAPP have not been successful due to its strong propensity to aggregate.

2 Systems and Methods

In this work, we performed MD computer simulation studies of the structural and hydrational properties of a single hIAPP peptide in liquid water in the temperature range from 250 to 450 K. All atomic molecular dynamics simulations were carried out with GRO-MACS v.3.3.1 using the OPLS-AA/L force field for the peptide and SPCE water molecules. Initially, the peptide was prepared in various starting conformations, including an α-helical conformation, four random conformations obtained from 1 ns runs at 1000 K *in vacuo*, of which one of the initial conformations being a fully extended isolated β-strand. After 15 to 30 ns simulation runs in water, the conformational behavior of hIAPP no longer depended on the initial configuration used. After 50 ns of equilibration at each temperature studied,

153

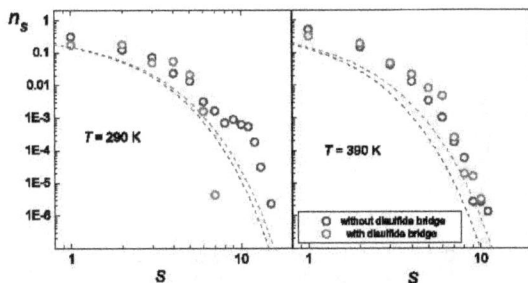

Figure 1. Probability n_S to find S successive residues with helical conformation. The dashed lines show n_S for a random distribution of residues in an infinite chain with the same content p of residues with analogous structure: $n_S = (1 - p)^2 p^S$.

200 ns trajectories were used for the analysis of the system properties. Two moieties of hIAPP were studied: hIAPP with and without the natural disulfide bridge between C2 and C7 residues.

3 Structural Properties

Analysis of the secondary structure shows that at all the temperatures studied, hIAPP does not adopt a well-defined conformation. The helical content of hIAPP, estimated as a fraction of residues having the dihedral angles within the allowed region of the Ramachandran plot, do not depend noticeably on the presence of a disulfide bridge and decrease upon heating. However, the ability of the helical residues to form a continuous sequence along the peptide chain is strongly suppressed by the disulfide bridge. This can be seen from the comparison of the probability distributions n_S to find S successive residues with helical conformation shown in Fig. 1. Large clusters of residues with helical dihedral angles disappear by introducing the disulfide bridge and by heating.

4 Volumetric Properties

The intrinsic volumetric properties of a biomolecule in water can be studied, when the density of hydration water is known.[1] The temperature dependence of the density ρ_h of the hydration water in a shell 0.3 nm thick at the hIAPP surface and of the density ρ_b of a bulk liquid water are shown in Fig. 2: ρ_h is below ρ_b and its temperature dependence is essentially linear. The temperature dependence of the density of hydration water indicates that the effective hydrophobicity of the hIAPP surface is close to that of carbon-like surfaces. Knowing the temperature dependences of ρ_h and ρ_b, we can estimate the intrinsic volume V_{int} of hIAPP from the equation: $V_{int} = V_{app} - \Delta V$. Here, V_{app} is the apparent volume of hIAPP measured as the difference between the volumes of the simulation boxes

154

113

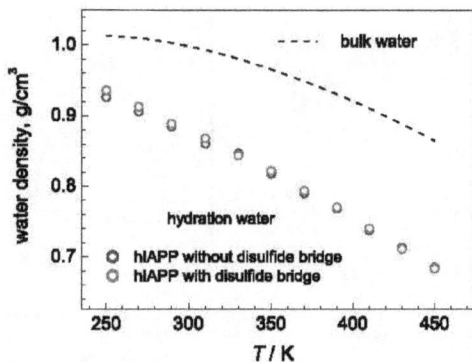

Figure 2. Temperature dependence of the density of bulk water and hydration water near hIAPP.

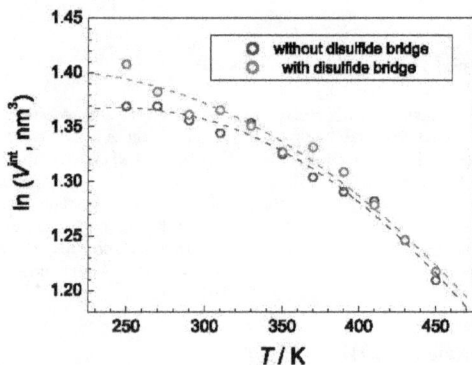

Figure 3. The temperature dependence of the logarithm of V_{int} of hIAPP. The lines are the fits to a quadratic polynomial.

with and without hIAPP, respectively, both having the same number of water molecules. The term ΔV accounts for the change of the system volume due to the different densities of hydration and bulk water, $\Delta V = V_h(1 - \rho_h/\rho_b)$, where V_h is the volume of hydration water. In a first approximation, V_h is the product of the solvent accessible area and the thickness of the hydration shell. The temperature dependence of the logarithm of V_{int} of

155

114

hIAPP is shown in Fig. 3. The slope of this dependence is equal to the intrinsic thermal expansion coefficient α_{int}. Similarly to the case of amyloid β peptide (1-42)[1], α_{int} of hIAPP is negative and becomes even more negative upon heating. Such behavior can be attributed to a decreasing helical content and a decreasing number of intrapeptide H-bonds. Note, that the disintegration of large clusters of helical residues by the disulfide bridge at low temperature (see Fig. 1) makes α_{int} more negative (see Fig. 3).

5 Thermal Disruption of the Hydration Water Network at the hIAPP Surface

The spanning H-bonded network of hydration water, which covers hIAPP homogeneously at low temperatures, breaks via a quasi-2D percolation transition, whose midpoint is located at about 320 K. Interestingly, approximately at this temperature, the experimentally measured lag time of hIAPP aggregation drops drastically[2]. Hence, we might conclude that the breakdown of the spanning H-Bonding network of hydration water might foster hIAPP aggregation.

Acknowledgments

Financial support from the International Max-Planck Research School in Chemical Biology and from the Federal State of NRW and the EU (Europäischer Fonds für regionale Entwicklung) is gratefully acknowledged.

References

1. I. Brovchenko, R. R. Burri, A. Krukau, A. Oleinikova, and R. Winter, *Intrinsic thermal expansivity and hydrational properties of amyloid peptide Aβ42 in liquid water*, to be published.
2. R. Kayed *et al.*, *Conformational transitions of islet amyloid polypeptide (IAPP) in amyloid formation in vitro*, J. Mol. Biol., **287**, 781, 1999.

Appendix B

Poster Presentations

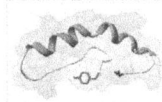
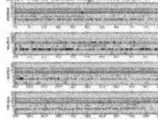
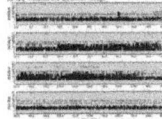
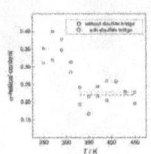
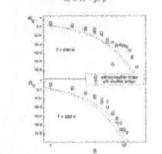
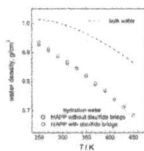
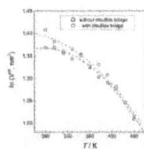
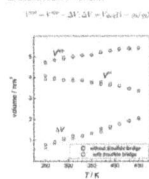
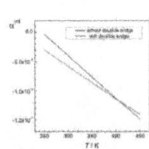
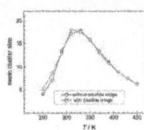
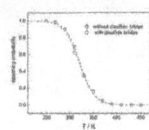

Figure B.1: *Poster presented at "From Computational Biophysics to Systems Biology (CBSB08)", Jülich, Germany.*

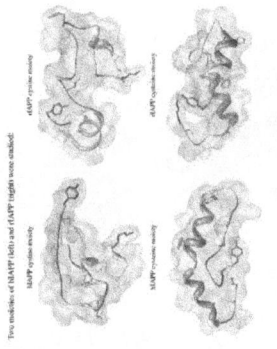
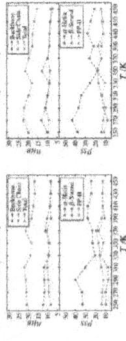
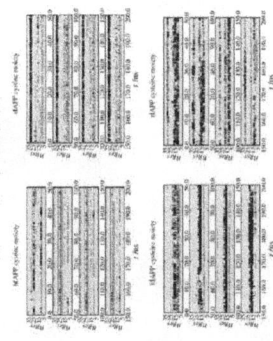
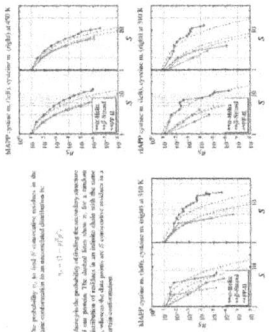

Figure B.2: *Poster presented at the 53rd Annual Meeting of the Biophysical Society, Boston, Massachusetts and at "Amyloid 2009," Halle (Saale), Germany.*

Bibliography

[1] Stryer, L., 1996. Biochemistry. W. H. Freeman and Company, 4th edition. Trans. Pessino, A., Bianca Sparatore, B. Rev. Melloni, E.

[2] Schlick, T., 2006. Molecular Modeling and Simulation: An Interdisciplinary Guide. Springer, New York.

[3] Meldi, D., 2004. Dizionario Etimologico. Rusconi Libri SrL, Santarcangelo di R. (Rn).

[4] Bergmann, M., 2004. Schülerduden, Chemie: [ein Lexikon Zum Chemieunterricht; Das Grundwissen Zur Chemie, Zu Ihren Begriffen, Verfahren Und Gesetzen; Aktuell - Kompetent - Verständlich Von A Bis Z]. Dudenverl., Mannheim; Leipzig; Wien; Zurich.

[5] Creighton, T. E., 1993. Proteins: Structure and Molecular Properties. W. H. Freeman and Company, New York, 2nd edition. 8th reprint 2006.

[6] Lodish, H., A. Berk, S. L. Zipursky, P. Matsudaira, D. Baltimore, and J. E. Darnell, 2000. Molecolar Cell Biology. W. H. Freeman and Company. 4th printing 2001.

[7] Kapurniotu, A., 2001. Amyloidogenicity and cytotoxicity of islet amyloid polypeptide. *Biopolymers* 60:438–459.

[8] Lopes, D. H. J., A. Meister, A. Gohlke, A. Hauser, A. Blume, and R. Winter, 2007. Mechanism of islet amyloid polypeptide fibrillation at lipid interfaces studied by infrared reflection absorption spectroscopy. *Biophys. J.* 93:3132–3141.

[9] Jha, S., D. Sellin, R. Seidel, and R. Winter, 2009. Amyloidogenic propensities and conformational properties of ProIAPP and IAPP in the presence of lipid bilayer membranes. *J. Mol. Biol.* 389:907–920.

[10] Kayed, R., J. Bernhagen, N. Greenfield, K. Sweimeh, H. Brunner, W. Voelter, and A. Kapurniotu, 1999. Conformational transitions of islet amyloid polypeptide (IAPP) in amyloid formation in vitro. *J. Mol. Biol.* 287:781–796.

[11] LeRoith, D., S. I. Taylor, and J. M. Olefsky, 2004. Diabetes Mellitus: A Fundamental and Clinical Text. Lippincott, Williams, and Wilkins, Philadelphia. Chapter 10.

[12] Uversky, V. N., and A. L. Fink, 2006. Protein Misfolding, Aggregation, and Conformational Diseases. Springer, New York. Chapter 10.

[13] Westermark, P., U. Engstrï£¡m, K. H. Johnson, G. T. Westermark, and C. Betsholtz, 1990. Islet amyloid polypeptide: pinpointing amino acid residues linked to amyloid fibril formation. *Proc. Natl. Acad. Sci. U. S. A.* 87:5036–5040.

[14] Green, J., C. Goldsbury, T. Mini, S. Sunderji, P. Frey, J. Kistler, G. Cooper, and U. Aebi, 2003. Full-length rat amylin forms fibrils following substitution of single residues from human amylin. *J. Mol. Biol.* 326:1147–1156.

[15] Porte, D., R. S. Sherwin, A. Baron, M. Ellenberg, and H. Rifkin, 2003. Ellenberg and Rifkin's Diabetes Mellitus: Theory and Practice. McGraw-Hill, Health Professions Division, New York.

[16] Lewis, B., 2000. Genes VII. Oxford University Press Inc., New York. 2nd reprint.

[17] Voet, D., and J. G. Voet, 1993. Biochimica. Zanichelli, Bologna. Trans. Giorgio Corte. Rev. Giorgio.Corte. 4th reprint 2002.

[18] Sakagashira, S., H. J. Hiddinga, K. Tateishi, T. Sanke, T. Hanabusa, K. Nanjo, and N. L. Eberhardt, 2000. S20G mutant amylin exhibits increased in vitro amyloidogenicity and increased intracellular cytotoxicity compared to wild-type amylin. *Am. J. Pathol.* 157:2101–2109.

[19] Fersht, A. R., and V. Daggett, 2002. Protein folding and unfolding at atomic resolution. *Cell* 108:573–582.

[20] Dayhoff, M. O., L. T. Hunt, W. C. Barker, R. M. Schwartz, and B. C. Orcutt, 1979. Atlas of Protein Sequence and Structure, volume 5 suppl. 3. National Biomedical Research Foundation, Washington D.C.

[21] Williamson, J. A., and A. D. Miranker, 2007. Direct detection of transient alpha-helical states in islet amyloid polypeptide. *Protein Sci.* 16:110–117.

[22] Brovchenko, I., R. R. Burri, A. Krukau, and A. Oleinikova, 2009. Thermal expansivity of amyloid beta(16-22) peptides and their aggregates in water. *Phys. Chem. Chem. Phys.* 11:5035–5040.

[23] Oleinikova, A., and I. Brovchenko, 2006. Percolation transition of hydration water in biosystems. *Mol. Phys.* 104:3841–3855.

[24] Oleinikova, A., and I. Brovchenko, 2006. Percolating networks and liquid-liquid transitions in supercooled water. *J. Phys.: Condens. Matt.* 18:S2247–S2259.

[25] Higham, C. E., E. T. Jaikaran, P. E. Fraser, M. Gross, and A. Clark, 2000. Preparation of synthetic human islet amyloid polypeptide (IAPP) in a stable conformation to enable study of conversion to amyloid-like fibrils. *FEBS Lett.* 470:55–60.

[26] Goldsbury, C., K. Goldie, J. Pellaud, J. Seelig, P. Frey, S. A. Mï£¡ller, J. Kistler, G. J. Cooper, and U. Aebi, 2000. Amyloid fibril formation from full-length and fragments of amylin. *J. Struct. Biol.* 130:352–362.

[27] Padrick, S. B., and A. D. Miranker, 2001. Islet amyloid polypeptide: identification of long-range contacts and local order on the fibrillogenesis pathway. *J. Mol. Biol.* 308:783–794.

[28] Mishra, R., M. Geyer, and R. Winter, 2009. NMR spectroscopic investigation of early events in IAPP amyloid fibril formation. *ChemBioChem* 10:1769–1772.

[29] Knight, J. D., J. A. Hebda, and A. D. Miranker, 2006. Conserved and cooperative assembly of membrane-bound alpha-helical states of islet amyloid polypeptide. *Biochemistry* 45:9496–9508.

[30] Flory, P. J., 1969. Statistical Mechanics of Chain Molecules. Wiley, New York.

[31] Li, S. C., N. K. Goto, K. A. Williams, and C. M. Deber, 1996. Alpha-helical, but not beta-sheet, propensity of proline is determined by peptide environment. *Proc. Natl. Acad. Sci. U. S. A.* 93:6676–6681.

[32] Richardson, J. S., and D. C. Richardson, 1988. Amino acid preferences for specific locations at the ends of alpha helices. *Science* 240:1648–1652.

[33] Nilsson, M. R., and D. P. Raleigh, 1999. Analysis of amylin cleavage products provides new insights into the amyloidogenic region of human amylin. *J. Mol. Biol.* 294:1375–1385.

[34] Koo, B. W., and A. D. Miranker, 2005. Contribution of the intrinsic disulfide to the assembly mechanism of islet amyloid. *Protein Sci.* 14:231–239.

[35] Vaiana, S. M., R. B. Best, W.-M. Yau, W. A. Eaton, and J. Hofrichter, 2009. Evidence for a partially structured state of the amylin monomer. *Biophys. J.* 97:2948–2957.

[36] Dupuis, N. F., C. Wu, J.-E. Shea, and M. T. Bowers, 2009. Human islet amyloid polypeptide monomers form ordered beta-hairpins: a possible direct amyloidogenic precursor. *J. Am. Chem. Soc.* 131:18283–18292.

[37] Lindahl, E., B. Hess, and D. van der Spoel, 2001. GROMACS 3.0: A package for molecular simulation and trajectory analysis. *J. Mol. Model.* 7:306–317.

[38] Berendsen, H. J. C., D. van der Spoel, and R. van Drunen, 1995. GROMACS: A message-passing parallel molecular dynamics implementation. *Comp. Phys. Comm.* 91:43–56.

[39] van der Spoel, D., E. Lindahl, B. Hess, G. Groenhof, A. Mark, and H. J. C. Berendsen, 2005. GROMACS: Fast, Flexible and Free. *J. Comput. Chem.* 26:1701–1719.

[40] Brovchenko, I., A. Krukau, N. Smolin, A. Oleinikova, A. Geiger, and R. Winter, 2005. Thermal breaking of spanning water networks in the hydration shell of proteins. *J. Chem. Phys.* 123:224905.

[41] Krukau, A., I. Brovchenko, and A. Geiger, 2007.   Temperature-induced conformational transition of a model elastin-like peptide GVG(VPGVG)$_3$ in water. *Biomacromolecules* 8:2196–2202.

[42] Oleinikova, A., I. Brovchenko, and G. Singh, 2010.   The temperature dependence of the heat capacity of hydration water near biosurfaces from molecular simulations. *EPL* 90:36001.

[43] Hansmann, U. H. E., J. H. Meinke, S. Mohanty, W. Nadler, and O. Zimmermann, editors, 2008.   From Computational Biophysics to Systems Biology (CBSB08), volume 40 of *NIC series*. NIC-Directors.

[44] Andrews, M. N., and R. Winter, 2010.   Comparing the structural properties of human and rat islet amyloid polypeptide by md computer simulations. *Biophys. Chem.* in print.

[45] Leach, A. R., 2001.   Molecular Modelling: Principles and Applications. Prentice Hall, Harlow, 2nd edition.

[46] Schaftenaar, G., and J. Noordik, 2000.   Molden: a pre- and post-processing program for molecular and electronic structures. *J. Comput. Aided Mol. Des.* 14:123–134.

[47] Allen, F. H., O. Kennard, D. G. Watson, L. Brammer, A. G. Orpen, and R. Taylor, 1987.   Tables of bond lengths determined by x-ray and neutron diffraction. part 1. bond length in organic compounds. *J. Chem. Soc., Perkin Trans.* 2 S1–19.

[48] Guex, N., and M. Peitsch, 1997. SWISS-MODEL and the Swiss-PdbViewer: An environment for comparative protein modeling. *Electrophoresis* 18:2714–2723.

[49] Jorgensen, W. L., and J. Tirado-Rives, 1988.   The OPLS [optimized potentials for liquid simulations] potential functions for proteins, energy minimizations for crystals of cyclic peptides and crambin. *J. Am. Chem. Soc.* 110:1657–1666.

[50] Kaminski, G., R. Friesner, J. Tirado-Rives, and W. Jorgensen, 2001. Evaluation and reparametrization of the OPLS-AA force field for proteins via comparison with accurate quantum chemical calculations on peptides. *J. Phys. Chem. B* 105:6474–6487.

[51] Dawson, R., D. Elliot, W. Elliot, and K. Jones, 1969.   Data for biochemical research. Oxford University Press, 2nd edition.

[52] Donnini, S., F. Tegeler, G. Groenhof, and H. Grubmueller, 2009.   Constant pH simulations in explicit solvent using the lambda-dynamics approach. *Biophys. J.* 96:574a.

[53] Liu, D. C., and J. Nocedal, 1989.   On the limited memory BFGS method for large scale optimization. *Math. Program. B* 45:503–528.

[54] Zimmerman, K., 1991.   All purpose molecular mechanics simulator and energy minimizer. *J. Comput. Chem.* 12:310–319.

[55] Essmann, U., L. Perera, M. L. Berkowitz, T. Darden, H. Lee, and L. G. Pedersen, 1995. A smooth particle mesh Ewald method. *J. Chem. Phys.* 103:8577–8593.

[56] Berendsen, H. J. C., J. Postma, A. DiNola, and J. Haak, 1984. Molecular dynamics with coupling to an external bath. *J. Chem. Phys.* 81:3684–3690.

[57] Garcia, A. E., 2004. Characterization of non-alpha helical conformations in Ala peptides. *Polymer* 45:669–676.

[58] Nosé, S., 1984. A molecular dynamics method for simulations in the canonical ensemble. *Mol. Phys.* 52:255–268.

[59] Hoover, W., 1985. Canonical dynamics: equilibrium phase-space distributions. *Phys. Rev. A* 31:1695–1697.

[60] Berendsen, H. J. C., J. R. Grigera, and T. P. Straatsma, 1987. The missing term in effective pair potentials. *J. Phys. Chem.* 91:6269–6271.

[61] Parrinello, M., and A. Rahman, 1981. Polymorphic transitions in single crystals: A new molecular dynamics method. *J. Appl. Phys.* 52:7182–7190.

[62] Nosé, S., and M. Klein, 1983. Constant pressure molecular dynamics for molecular systems. *Mol. Phys.* 50:1055–1076.

[63] Miyamoto, S., and P. Kollman, 1992. Settle: An analytical version of the SHAKE and RATTLE algorithms for rigid water models. *J. Comput. Chem.* 12:952–962.

[64] Ryckaert, J., G. Ciccotti, and H. J. C. Berendsen, 1977. Numerical integration of the cartesian equations of motion of a system with constraints; molecular dynamics of n-alkanes. *J. Comput. Phys.* 23:327–341.

[65] Darden, T., D. York, and L. Pedersen, 1993. Particle mesh Ewald: An N-log(N) method for Ewald sums in large systems. *J. Chem. Phys.* 98:1311–1327.

[66] Oleinikova, A., N. Smolin, I. Brovchenko, A. Geiger, and R. Winter, 2005. Formation of spanning water networks on protein surfaces via 2D percolation transition. *J. Phys. Chem. B* 109:1988–1998.

[67] Brovchenko, I., R. R. Burri, A. Krukau, A. Oleinikova, and R. Winter, 2008. Intrinsic thermal expansivity and hydrational properties of amyloid peptide abeta42 in liquid water. *J. Chem. Phys.* 129:195101.

[68] Eisenberg, D., and A. McLachlan, 1986. Solvation energy in protein folding and binding. *Nature* 319:199–203.

[69] D. van der Spoel, B. H., E. Lindahl, 2005. Gromacs User Manual version 3.3.

[70] Taylor, J. R., 1982. An Introduction to Error Analysis, The Study of Uncertainties in Physical Measurements. Univeristy Science Books, 1st edition. Trans. Caporaloni, M. Rev. Palmonari, F.

[71] Hornak, V., R. Abel, A. Okur, B. Strockbine, A. Roitberg, and C. Simmerling, 2006. Comparison of multiple Amber force fields and development of improved protein backbone parameters. *Proteins* 65:712–725.

[72] Bland, J. M., and D. G. Altman, 1986. Statistical methods for assessing agreement between two methods of clinical measurement. *Lancet* 1:307–310.

[73] Kahn, K., and T. C. Bruice, 2002. Parameterization of OPLS-AA force field for the conformational analysis of macrocyclic polyketides. *J. Comput. Chem.* 23:977–996.

[74] Ramachandran, G. N., and V. Sasisekharan, 1968. Conformation of polypeptides and proteins. *Adv. Protein Chem.* 23:283–438.

[75] IUPAC-IUB Commission on Biochemical Nomenclature, 1970. Abbreviations and symbols for the description of the conformation of polypeptide chains. tentative rules (1969). *Biochemistry* 9:3471–3479. IUPAC-IUB Commission on Biochemical Nomenclature.

[76] Seeliger, D., 2004. Python/Gromacs Interface.

[77] Kabsch, W., and C. Sander, 1983. Dictionary of protein secondary structure: pattern recognition of hydrogen-bonded and geometrical features. *Biopolymers* 22:2577–2637.

[78] Andersen, C. A. F., A. G. Palmer, S. Brunak, and B. Rost, 2002. Continuum secondary structure captures protein flexibility. *Structure* 10:175–184.

[79] Cubellis, M. V., F. Cailliez, and S. C. Lovell, 2005. Secondary structure assignment that accurately reflects physical and evolutionary characteristics. *BMC Bioinformatics* 6 Suppl 4:S8.

[80] Hess, B., 2002. Determining the shear viscosity of model liquids from molecular dynamics simulations. *J. Chem. Phys.* 116:209–217.

[81] Park, J., S. A. Teichmann, T. Hubbard, and C. Chothia, 1997. Intermediate sequences increase the detection of homology between sequences. *J. Mol. Biol.* 273:349–354.

[82] Kuster, D. J., S. Urahata, J. W. Ponder, and G. R. Marshall, 2009. From data or dogma? The myth of the ideal helix. *Biophys. J.* 96:5a.

[83] Alejandre, J., D. J. Tildesley, and G. A. Chapela, 1995. Molecular dynamics simulation of the orthobaric densities and surface tension of water. *J. Chem. Phys.* 102:4574–4583.

[84] Lide, D. R., 1995. Handbook of Chemistry and Physics. CRC Press, Boca Raton, Florida. 76th edition.

[85] Báez, L. A., and P. Clancy, 1994. Existence of a density maximum in SPC/E-modeled water. *J. Chem. Phys.* 101:9837–9840.

[86] Kirkup, L., 1994. Experimental Methods: An Introduction to the Analysis and Presentation of Data. Wiley, Australia.

[87] Allen, M. P., and D. J. Tildesley, 1989.   Computer Simulation of Liquids. Calerdon Press, Oxford.

[88] Savitzky, A., and M. J. E. Golay, 1964. Smoothing and differentiation of data by simplified least squares procedures. *Anal. Chem.* 36:1627–1639.

[89] Scott, D. W., 1979.   On optimal and data-based histograms.   *Biometrika* 66:605–610.

[90] Oleinikova, A., I. Brovchenko, N. Smolin, A. Krukau, A. Geiger, and R. Winter, 2005. Percolation transition of hydration water: from planar hydrophilic surfaces to proteins. *Phys. Rev. Lett.* 95:247802.

[91] Daura, X., K. Gademann, J. Bernhard, D. Seebach, W. F. van Gunsteren, and A. E. Mark, 1999. Peptide folding: When simulation meets experiment. *Angew. Chem. Int. Ed. Engl.* 38:236–240.

[92] Mayor, U., C. M. Johnson, V. Daggett, and A. R. Fersht, 2000. Protein folding and unfolding in microseconds to nanoseconds by experiment and simulation. *Proc. Natl. Acad. Sci. U. S. A.* 97:13518–13522.

[93] Ferguson, N., J. R. Pires, F. Toepert, C. M. Johnson, Y. P. Pan, R. Volkmer-Engert, J. Schneider-Mergener, V. Daggett, H. Oschkinat, and A. Fersht, 2001. Using flexible loop mimetics to extend phi-value analysis to secondary structure interactions. *Proc. Natl. Acad. Sci. U. S. A.* 98:13008–13013.

[94] Thornton, J. M., and B. L. Sibanda, 1983. Amino and carboxy-terminal regions in globular proteins. *J Mol Biol* 167:443–460.

[95] Yonemoto, I. T., G. J. A. Kroon, H. J. Dyson, W. E. Balch, and J. W. Kelly, 2008. Amylin proprotein processing generates progressively more amyloidogenic peptides that initially sample the helical state. *Biochemistry* 47:9900–9910.

[96] Kumar, S., and R. Nussinov, 2002.   Close-range electrostatic interactions in proteins. *Chembiochem* 3:604–617.

[97] Langley, R., 1971. Practical statistics simply explained. Dover Publications, New York, revised edition.

[98] Theil, H., 1961. Economic Forecasts and Policy. North-Holland Publishing Company, Amsterdam.

[99] Brovchenko, I., and A. Oleinikova, 2008.   Interfacial and Confined Water. Elsevier, Amsterdam.

[100] Brovchenko, I., A. Geiger, and A. Oleinikova, 2004. Water in nanopores: II. The liquid-vapour phase transition near hydrophobic surfaces. *J. Phys.: Condens. Matt.* 16:S5345–S5370.

[101] Brovchenko, I., and A. Oleinikova, 2006. Surface critical behaviour of fluids and magnets. *Mol. Phys.* 104:3535–3549.

[102] Oleinikova, A., I. Brovchenko, and R. Winter, 2009. Volumetric properties of hydration water. *J. Phys. Chem. C* 113:11110–11118.

[103] Berry, R. S., S. A. Rice, and J. Ross, 2000. Physical Chemistry. Oxford University Press, New York, 2nd edition.

[104] Barron, T. H. K., and G. K. White, 1999. Heat Capacity and Thermal Expansion at Low Temperatures. Kluwer Academic/Plenum, New York.

[105] Oleinikova, A., I. Brovchenko, and A. Geiger, 2006. Percolation transition of hydration water at hydrophilic surfaces. *Physica A* 364:1–12.

[106] Brovchenko, I., and A. Oleinikova, 2008. Which properties of a spanning network of hydration water enable biological functions? *ChemPhysChem* 9:2695–2702.

[107] Smolin, N., A. Oleinikova, I. Brovchenko, A. Geiger, and R. Winter, 2005. Properties of spanning water networks at protein surfaces. *J. Phys. Chem. B* 109:10995–11005.

[108] Tenidis, K., M. Waldner, J. Bernhagen, W. Fischle, M. Bergmann, M. Weber, M. L. Merkle, W. Voelter, H. Brunner, and A. Kapurniotu, 2000. Identification of a penta- and hexapeptide of islet amyloid polypeptide (IAPP) with amyloidogenic and cytotoxic properties. *J. Mol. Biol.* 295:1055–1071.

[109] Sterpone, F., C. Bertonati, G. Briganti, and S. Melchionna, 2009. Key role of proximal water in regulating thermostable proteins. *J. Phys. Chem. B* 113:131–137.

[110] Stauffer, D., and A. Aharony, 1992. Introduction to Percolation Theory. Taylor and Francis, London.

[111] Reynolds, P. J., H. E. Stanley, and W. Klein, 1977. Ghost fields, pair connectedness, and scaling: exact results in one-dimensional percolation. *J. Phys. A: Math. Gen.* 10:L203–L209.

[112] Brovchenko, I., M. N. Andrews, and A. Oleinikova, 2010. Volumetric properties of human islet amyloid polypeptide in liquid water. *Phys. Chem. Chem. Phys.* 12:4233–4238.

[113] Soong, R., J. R. Brender, P. M. Macdonald, and A. Ramamoorthy, 2009. Association of highly compact type II diabetes related islet amyloid polypeptide intermediate species at physiological temperature revealed by diffusion nmr spectroscopy. *J. Am. Chem. Soc.* 131:7079–7085.

[114] Marek, P., A. Abedini, B. Song, M. Kanungo, M. E. Johnson, R. Gupta, W. Zaman, S. S. Wong, and D. P. Raleigh, 2007. Aromatic interactions are not required for amyloid fibril formation by islet amyloid polypeptide but do influence the rate of fibril formation and fibril morphology. *Biochemistry* 46:3255–3261.

[115] Pettersen, E. F., T. D. Goddard, C. C. Huang, G. S. Couch, D. M. Greenblatt, E. C. Meng, and T. E. Ferrin, 2004. UCSF Chimera–a visualization system for exploratory research and analysis. *J. Comput. Chem.* 25:1605–1612.

[116] Moriarty, D. F., and D. P. Raleigh, 1999. Effects of sequential proline substitutions on amyloid formation by human amylin20-29. *Biochemistry* 38:1811–1818.

[117] Azriel, R., and E. Gazit, 2001. Analysis of the minimal amyloid-forming fragment of the islet amyloid polypeptide. An experimental support for the key role of the phenylalanine residue in amyloid formation. *J. Biol. Chem.* 276:34156–34161.

[118] Nanga, R. P. R., J. R. Brender, J. Xu, K. Hartman, V. Subramanian, and A. Ramamoorthy, 2009. Three-dimensional structure and orientation of rat islet amyloid polypeptide protein in a membrane environment by solution nmr spectroscopy. *J. Am. Chem. Soc.* 131:8252–8261.

[119] Persistence of Vision Pty. Ltd., 2004. Persistence of Vision Raytracer (Version 3.6). Persistence of Vision Pty. Ltd.

[120] Sanner, M. F., A. J. Olson, and J. C. Spehner, 1996. Reduced surface: an efficient way to compute molecular surfaces. *Biopolymers* 38:305–320.

[121] Kyte, J., and R. Doolittle, 1982. A simple method for displaying the hydropathic character of a protein. *J. Mol. Biol.* 157:105.

[122] Vaiana, S. M., R. Ghirlando, W.-M. Yau, W. A. Eaton, and J. Hofrichter, 2008. Sedimentation studies on human amylin fail to detect low-molecular-weight oligomers. *Biophys. J.* 94:L45–L47.